Ballard, Edward G.

Man and technolo-
gy

*Quaestio mihi factus sum.*
St. Augustine

# MAN
## and
# TECHNOLOGY

Toward the Measurement

of a Culture

*by*

Edward Goodwin Ballard

DUQUESNE UNIVERSITY PRESS
PITTSBURGH
*Distributed by Humanities Press, Atlantic Highlands*

Second Printing, 1979

Library of Congress Cataloging in Publication Data

Ballard, Edward G,
    Man and technology.

Includes bibliographical references and index.
1. Ontology. 2. Civilization. Modern – 20th century. 3. Technology and ethics. I. Title.

BD311.B3                             111                    77-20902
ISBN-0-391-00751-3
ISBN-0-391-01048-4 pbk

Manufactured in the United States of America

FIRST EDITION

*For my son,*

Edward Marshall Ballard
1957-1974

# CONTENTS

# PREFACE

Anyone who thinks philosophically learns quickly to participate only ironically in his culture. The reason for this irony arises not merely from the fact that culture is for most of us an accident of birth but more profoundly from the fact that the institutions, principles, achievements and aims of any culture are ambiguous. Their ambiguity stands forth most evidently in valuations but is no less present in other contexts of thought or activity. For example, the time and energy consumed in investigating the nature of the electron could have been employed in other perhaps equally desirable projects. The consideration given to the writing of this book, as of any other, could have been usefully engaged in a different task. Generally speaking, any positive value is at the same time a negative value. A whole culture even has its darker side; and our accomplishments no less than our failures are haunted by what might have been. Philosophy can render us the important service of recalling and exhibiting to us the losses which our positive achievements have entailed; thus, it helps to prepare us for the ironies of the future.

The function of this book is to turn thoughtful consideration back upon elements of ourselves and of our possibility which we are in danger of leaving behind and forgetting as we move further in the direction of popular achievements and fashionable goals. Thus, this is a book which seeks to cultivate irony and to preserve a questioning attitude toward our present selves and our world. It is explicitly not sure that, as Shakespeare has Ophelia say in her madness, "We know what we are but know not what we may be." And it seeks to express the rationale of this uncertainty and the ground of this irony.

Not unnaturally this pursuit takes the form of a critique of culture, of twentieth century Westernized culture. If it can raise some provocative questions concerning ourselves and our relation to this culture and to this world, it will have done its job.

I am deeply obligated to Professors Charles Bigger, David Cornay, and Edward Henderson, of Louisiana State University, and to Professor Michael Zimmerman of Tulane University, who went far beyond the call of duty in reading the whole typescript, in making many cogent criticisms, and in offering useful, in some cases even indispensable, contributions. Professor Larry Miller of Tulane University read Chapter VI and provided a number of useful suggestions. I am deeply obligated to these gentlemen and to the many others who have helped me in this writing. The mistakes, obscurities, and confusions in this book are, of course, my own.

I want to thank editors and publishers for permissions to make such use as I needed of the following articles of mine: "Unmasking the Person," *Southern Journal of Philosophy,* vol. 11, nos. 1 and 2 (1973), pp. 7-14; "On the Phenomenon of Obligation," *Tulane Studies in Philosophy,* vol. 21, ed. Andrew J. Reck (The Hague: Martinus Nijhoff, 1972), pp. 139-157; "On Truth, its Nature, Context, and Source," *Man and World* vol. 1, no. 1 (1968), pp. 113-136; "Toward a Phenomenology of Man," *Proceedings of the American Catholic Association* (Washington, D.C.: The Catholic University of America, 1968), 169-174; "World and Culture," *Humanity and Technology,* ed. David Lovekin, to be published. And, finally, I want to thank Mrs. Freda Faber for preparation of the typescript.

*Tulane University, New Orleans, La.*

CHAPTER I

# OBJECTIVES AND BASIS

## 1. *Technology and Humanistic Philosophy*

To be technophobic in our time is to be willing to accept starvation or slavery. Yet the kind of technological culture to which present developments seem to be taking us may not be exactly what we intend. It may not be favorable to the human being either as he has hitherto known himself or as he envisages his future growth. Thus, men of our culture are urged by their situation in the world and in this epoch of history to take stock of themselves and of their present direction of development. This essay is intended to be a contribution to this stock taking.

The reigning view of man, of his future, and of his technology is, happily, not the only one. The purpose of this book is to elaborate an alternative understanding of man and his environment and to compare this understanding with a contemporary popular view. Thus, this book is in part polemical; in part it is descriptive and theoretical. The polemical part is developed in the form of a measurement of the technological culture toward which the Western countries seem to be developing. The descriptive and theoretical part will elaborate an account of man in his world which will provide a standard by which the measurement just indicated can be taken.[1]

In its polemical function, this essay will speak against a

1

certain view of science and technology in their relation to human life. The understanding of the nature of science and technology to which I refer can be characterized with reasonable accuracy as positivist and naturalistic. There is, to be sure, much to say for this view. One of the chief things to say for it is that it is widespread and influential. It counts among its proponents such men as these: Albert Einstein, P. W. Bridgeman, Jacques Monod, B. F. Skinner.[2] Popular as well as highly specialized pictures of the future of Western culture are deeply marked by the opinions and aspirations of these men and of others like them. Before this essay is ended, however, it will become evident that the opponent is more than this group of men and those who believe and sympathize with them.

The name which, with a great deal of trepidation, I shall assign to the opposing doctrine, to be developed in these pages, will be "humanism." Immediately it must be emphasized that the term is not used in any one of its usual senses, but perhaps its meaning here does come close to that which is common to many of its uses.[3] This common and basic meaning refers to the wonder aroused in us by the existence and fate of man and to the resolution to maintain consideration of this existence and fate at the center of concern.

Nothing recondite is intended by the main terms of this resolution. A thing may be said to exist if it manifests its own being (or could manifest it were there a being present able to perceive the manifestation). And a being is human if to some extent it can make its own being manifest both to itself and to others; this human self-manifestation is achieved mainly by means of symbols and language.[4] Such manifestation occurs within limits. Consideration of these limits can always be pressed, I believe, to the point where no further reasons can be found to account for their being just as they are. At this ultimate point an irrationality is encountered which may be appropriately termed "fate." These, in brief, are the meanings which I have in mind when I describe this essay as humanistic. It establishes concern for human being at the center of its interest, and it attempts to pursue this interest to its limits.

This essay will take for its polemical topic the evaluation of technological culture from the coign of vantage of the kind of humanism just indicated. And I repeat: such an evaluation is emphatically crucial at this time, when men appear to be deciding to turn their energies toward a more and more exclusive preoccupation with modern industrialized technology. For a decision in favor of such a culture will more than likely be irreversible. Techniques now being perfected in biology and psychology can so alter man's biological makeup and psychological climate that he will be deprived both of the possibility and the desire to change the direction of his evolution. Technology, in other words, is approaching the stage where embarking unreservedly upon its culture is a way of no return. Thus, our task is to develop an understanding of man, then to use this understanding as a standard in order to offer a humanistic measure of technological culture. Appropriate first steps will be to give some preliminary attention to the fundamentals of measuring and then to the basic ontological notions belonging to the context within which the present measurement is to be conducted.

## §2. *Two Kinds of Measure*

In the *Politicus* (284D-285A) Plato remarked that any art makes us call upon two sorts of measurement. There is, on one hand, the technique of measuring quantities, angles, length, time, weight, and the like, upon which the actual operation of even the manual arts—much less of the sciences—depends. Then there is also qualitative measurement of the excellence of the functioning of an art or of the adequacy with which an art satisfies the need that called it into being in the first place. Both kinds of measure, the art of quantitative measure and the art of measure by reference to a standard of excellence or adequacy, move rapidly upon being questioned into philosophic dimensions. Investigation of quantitative measure leads to problems concerning the nature of number, the relations obtaining between numbers and things, and the relations among ideas, concepts, and particular objects. Investigation of the excellence or adequacy of the product of an art leads into problems concerning the nature of value, the nature of art, the connections among the arts, the relation between arts and man, and the relation to the final good which the whole structure of his arts and man himself, may subserve.

Despite the keenness of his scent for a philosophic difficulty, Plato seems to have been less aware of the problematic relation between quantitative and qualitative measure than might have been expected. He let it go as more or less obvious that the arts, including the arts of quantitative measurement, are properly the means to human ends. Even mathematics, which occupies so large a place in the education of the ideal citizen (*Rep.* vii), is easily put down as a use value since it trains the mind to perceive ideas and to apprehend the intelligible order of the universe and of man. The final human good, which is the embodiment of this universal and intelligible harmony in one's own soul, is called upon to justify the exalted position which is assigned mathematical study in *paideia*.

To many philosophers, this view of the relation of mathematics and measurement to man appears to be simplistic. Upon second thought, however, it is the character of our own times rather than Plato's which should bring to the fore the complexity of the relation of mathematics as well as of science and technology to human life. For ours is the world which has carried the arts of quantitative measure to extraordinary lengths, and the technology which is founded upon skill in measuring has directed our desires quite away from the ideal of the heroic facing of one's fate, away from the attempt to embody a universal harmony, away from the hope of being an image of God, in fact away from any traditional view of the human good. In consequence, quantitative science, its mensuration, and its technology so dominate our thinking that the art of measuring by the standard of the good tends to appear to us now not to be respectable unless it too can be made quantitative. Hence, feverish efforts have been directed toward giving quantitative form to measurement by a standard of excellence. There are tests which are supposed to assign a numerical value to one's moral achievement, aesthetic sensitivity, personality adjustment, aggressiveness, friendliness, sociability, and many other human qualities. Are the claims made for these tests exaggerated? Are their results any good? The mathematics involved in calculating many such numerical values has been declared suspect.[5] But even if it were not suspect, there still remains the question of the use to which such quantitative techniques are put. Should they be used to help render the population more amenable to whatever requirements are prescribed by the present social order and its bureaucrats? Or should they be used as a means to the pursuit of healthful activity defined with reference to the good of man as man.

I wish to raise such questions as these on a still larger stage. Can the sciences themselves be measured? More particularly, can our culture, which increasingly is directing its major energies toward developing and applying the sciences and complacently places itself at the apex of evolution for this achievement, be measured? Is there such a thing as the human measure of technological culture? This last question will be best answered by the attempt actually to make the measurement.

This task, however, is easier to indicate than to accomplish. There is no man come from Mars bearing an absolute standard of value. Even if there were, we would be inclined to declare its absoluteness relative to Martian culture and then to raise difficult questions concerning the measure of Martian values. A further analysis of our own average standards or of our ordinary speech about them would fail utterly to reckon with the difficulty. For the Western world already manifests more than a few signs of becoming a thoroughly technological culture, and *ipso facto* would judge such a development to be desirable. Hence, a more original effort is demanded.

The general outlines of the task I indicate are fairly widely understood. For example, everyone in the modern world is familiar with the process of this sort of quantitative measurement (say, the measurement of temperature) and recognizes that some knowledge of the characteristics of the measuring instrument, of the object to be measured and of the operations of applying the former to the latter are necessary. The difference between my task and the quantitative illustration is that the quantitative one exists in a context which is already well ordered and quite familiar. Everyone knows that the ratio of the coefficient of expansion of mercury to that of glass is too small to be considered for ordinary purposes. In addition, every one knows or believes he knows how to apply and to read a thermometer. On the contrary, understanding of the nature of man, and of technology, and of the pertinent relations between them, decidedly does not exist in a common context well ordered and understood by all. Indeed, whether there can be any such common context is a debated point. The task, then, requires that these three items be developed from the beginning.

## §3. *Qualitative Measurement*

In matters of quantitative measurement, we know well enough how to proceed. As in the instance of the measure of heat intensity, once we know what is to be measured, we have only to agree upon a standard of measure and upon the operations and conditions for using it. But a qualitative or evaluative measure is less well understood. I suggest that up to a point we proceed analogously. We must first define that which is to be measured. Then this definition, describing the standard object, may be used over again as a measure. Such a procedure is used, for example, in the preparation of drugs for the pharmacist's use, in the specifications for various food products, in the preparation and regulation of a city's water supply. Consider the latter example: potable water is defined rather elaborately as $H_2O$ plus an acceptable level of various "impurities" and additives (e.g., disinfectants). It is defined thus by a specialist who is well acquainted with the chemistry of water as well as with the physiological needs and tolerances of the human organism. After treatment in the purification plant, samples of water passing through the conduits to the users are tested from time to time and the results of this analysis compared with the definition of potable water. Thus, the definition is used over again as a standard of measure. The comparison of the results of the analysis of samples with this standard is the measuring; this act enables the examiner to judge whether the water is good, just acceptable, or polluted. I offer this instance as a fair sample of qualitative measure. Qualitative measure, then, is the comparison of the object or state of affairs to be measured with the appropriate definition or description functioning as a standard of measure; this comparison provides the basis for a judgment of the degree to which the object or state of affairs embodies, fails to embody, or exceeds the definition or standard.

To take the measure of a culture as it passes through our hands to the next generation, would we not have to proceed in a manner rather similar to the physiological chemist's in the filtration plant? If so, then we would have to understand the nature of the environment and especially the nature of the man who makes and uses the culture, if he has a nature or a structure, and in this way reach an independent grasp of his possible well being. This understanding would serve as our definition and our standard of measure. Next, we would have to isolate and specify the significant elements of the culture in question and its view of man's nature. We would then have to compare these elements with the definition or standard of measure which in this instance would be our vision of man himself—were we to have such a veritable vision. Then we might decide by appeal to this comparison whether the culture is healthful, merely acceptable, or self-destructive as measured by this standard. This order is indeed a large one, but I shall count my efforts successful if they provide something for the reader to measure.

This order is a large one mainly because of the difficulty in defining man and thus in acquiring a standard of measure. Does man have a fixed and definable nature? It is certain that there are in modern times no generally received beliefs about the nature of man. Many philosophers and scientists would readily agree that we know more about the electron or about the galactic system than about man. Ancient times seemed to fare better although in Plato's *Parmenides* the young Socrates expressed a doubt that there was a form or definable nature of man, still in other dialogues (e.g., the *Republic*) Plato is quite clear about the soul and its appropriate care, and Aristotle was confident that a man was a structure of potentialities which one could and should actualize in quite definite ways. This serene confidence was lost, however, in the Christian world. The man of this world was a being who had fallen from his nature. His only hope for restoration and self-understanding was linked, according to orthodox doctrine, to a gift from a transcendent God accessible only through an institution. This institution, however, proved to be prey to all the human weaknesses. In modern times the account of his nature varies from the several remnants or adaptations of earlier humanistic and theological views to such typically recent additions as these: man is an ambulatory computer (sometimes oddly programmed), or a being who can learn to use the first person pronoun correctly, or a product of socio-economic processes, or a useless passion unpredictably free but enmeshed in a web of meaningless routine. This screed of discordant beliefs suggests that man may have no determinate nature. One of the problems which will have to be faced later is this: if man does indeed have no fixed or permanent nature, then he has no definition; but if he has no definition, then there is no measure of him or his culture of the sort indicated in the preceding two paragraphs. Such a conclusion seems to invite chaos into human affairs and explains well enough why mankind has clung tenaciously to traditional convictions. In the chapters to follow, we shall have to face this difficulty and attempt in spite of it to produce a qualitative standard for the measure of technological man and his culture.

The classic moral and humanistic tradition has guided men successfully through many centuries of tumultuous history and seems still to have much to recommend it. Why has it lost its position of leadership? One answer is that the important and once functional elements of the tradition have failed to deal adequately with the new realms opened up by the modern sciences and their applications. Traditional Western man did not foresee and was in no way prepared to deal effectively with the innovations and crises which have followed hard upon the birth and growth of modern science and its technological applications. Especially was he not prepared for the consequences of the application of the concepts and techniques of the sciences to the study and control of man himself. Several reactions to this failure have become apparent. Some intellectuals simply abandoned the tradition and faithfully expected either that science would develop its own wisdom or else that mankind would muddle through somehow until a Golden Age was reached. As it were, men of this faith persuaded themselves that continued evidence of technological progress is a sufficient measure of the healthful state of our culture. Others, reflecting upon changes occurring in modern times, draw a distinction between industrial man, the type most evident in the present society, who uses modern technology for egotistic ends, and technological man, who may emerge in the future, and who will use the technological instrument for the benefit of all. Industrial civilization, as we well know, has thrived on a heedless use of nature in a program of large scale production and consumption of material commodities, within an atmosphere of competition and war. Thus, it has endowed a minority with riches and power while continuing to restrain a large population at the level of comparative poverty and impotence. That which it chiefly values, riches and power, it has never been able to distribute justly. Is there any reason to believe that technological man, with a still more powerful instrument of self aggrandizement, will be any different? One writer, Victor C. Ferkiss, seems to believe that technological man will not only be different, but he will be better.[6] Just how he will become better, this author does not make clear. But the optimistic belief seems to be that by some process of progressive conditioning, and developing symbiosis with the machine, a new man—or man-machine—will arise who will bring humanity to a new and higher stage of existence.[7] At this stage, men will be in complete control of their environment, their society, and their own evolution. Thereupon, men will dispose of technological power in a completely "rational" manner to serve the public good. But this kind of blandly confident millenial utopianism has by now worn somewhat threadbare. Why should we expect technological man to be anything more than a more powerful and more efficient industrial man? After all, the child is father to the man. Perceptive critics of our times have seldom found the cure for our ills to lie in more of the same fare which produced the ills.

Men like Kierkegaard and Nietzsche directed radical research into the modern world and its weaknesses. Both repudiated technology and the mass culture which went with it. Kierkegaard also uncovered the death of religion in our time, and Nietzsche publicized the death of God. The one recognized the end of the means to our salvation, and the other recognized the end of this end. Together they bring us to the beginning of the technological epoch, and we are left "as on a darkling plane . . . " which those who have abandoned the tradition seek to illuminate with mercury vapor lamps and other gadgets of industry.

But the gadgets of technology are not just funny. At the moment of this writing, the United States government diverts something on the order of one quarter of its income to military uses. One product of this expenditure is an operative nuclear submarine fleet capable of delivering 7000 warheads each four times as powerful as the one that destroyed 100,000 persons in Hiroshima. Also three other nations possess such fleets. Those who like to take the measure of the effectiveness of such weapons speak with equanimity of the "overkill" they might cause among the enemy (whoever "the enemy" happens to be that year) and of the "biological discomfort" which would be produced by fallout among their own population. After what manner do those responsible for judgments such as these think of themselves and of their fellow man? It is not difficult to imagine them ordering brain operations for "troublesome" American soldiers in Vietnam merely in order to render these soldiers more tractable, or ordering some prisoners in an

Alabama penitentiary, without the prisoners' knowledge or informed consent, to be divided into two groups, one to take an experimental drug for the treatment of syphilis, the other, the control group, to take only a placebo.[8] These examples are illustrations of what those now in control tend to think of as the public good. Has man become a guinea pig to himself? Are the critics right? Is Western man well along toward losing his self-understanding? But here I seem to be giving way to moral indignation. I believe it has been correctly observed that those who indulge in moral indignation seldom do anything else. The remainder of this book will be devoted to something else. In short, the position will be taken that a rethinking of Western humanism since the Middle Ages and of contemporary thought about man, his technology, and his values is a very pressing necessity. Else we may take an irremediably wrong turn in the maze of history.

## §4. *The Ontological Basis and the Orientation of this Essay*

The ontological foundation for this rethinking will be presented only briefly. This essay is not a treatise on fundamentals; still some understanding of the basis of the present criticism and measurement of technological culture will facilitate grasp of the position to be developed. In this section I intend to provide an account of being, of perception, of experience, and of world which will be sufficient to enable us to proceed to the main task with proper grounding. The ontological topic will be broached appropriately by recalling an ancient myth. No doubt the further we go back in time as well as toward the basis of thought, the more similar do the elements of human experience become and the closer we may approach a basis which is common both to the technological and to the (in our sense) humanistic.

### A) BEING

A cosmogonic myth often encountered among primitive people, instanced among the ancient Greeks as well as among other peoples, tells that in the beginning our first ancestors, Earth and Sky, lay in a generative embrace.[9] The creatures to which they gave birth were cramped and had no room to move and breathe. So, the myths say, they united their strengths and somehow pushed Earth and Sky apart, forming thus "chaos"[10] or the world in which they could perceive and act. The first beings in this living space were born of Earth. Hesiod (*Theogony*) mentions several of her children: Eros, the Titans, eventually Zeus, the later sky god, who triumphed over the earth-born Giants and apportioned the universe out among his supporters. (No doubt the way in which the primal parents were separated had a great deal to do with the kind of open space for living made available to their progeny and with the kind of creatures their progeny became. And so it continues to be; the way in which we push aside our own generating forebears and inherited customs in order to gain some freedom and to make a culture for ourselves, determines the primary structure of our existence.)

In the myth the original being, Ouranos-Gaea, seemingly one, was in fact complex and potentially structured. Its continuation or its persisting in being itself consisted in this complexity becoming explicit. The first statement moving out of myth and initiating an ontology points toward this fundamental structure of being. The most fruitful statement indicative of this peculiar complexity which I know is offered by Plato in the *Sophist*. In the context of the great battle between the materialistic Giants and the peaceable Friends of the Forms, Plato describes being as the power to affect or to be affected (247D). Note that being is not merely power or possibility, nor is it merely the directional factor which renders power effective. Always it is both. Being manifests its presence by affecting and being affected in determinate ways.

This statement about being and the myth are mutually interpreting. Both, likewise, are reflected in a familiar story about pre-history. We remember that that strange being, early man, the myth-maker, sometimes lived in fertile climes and places. Here he was the food gatherer, attached to the soil which he regarded as a mother, nourishing and protecting him. His was a religion of earth, of chthonic goddesses and fertility rites. But sometimes he lived in mountains and relatively arid places. There he became the hunter, following the food providing animals, and locating himself under the sky and by means of its luminaries. The religion of these nomadic people was centered upon the sky gods, strong beings skilled in the arts of war and the hunt. When these toughened hunters came upon and took over a fertile area and its peasant population, they became intensely conscious of boundaries. The Doric tribes, for example, who came down from the North upon the farming peoples around the Greek peninsula, brought with them their sky gods. These gods came to dominate, at least for a time, the Titans and earth deities whom they found already at hand. Their chief god, Zeus, parcelled out the cosmos to his adhering divinities, Poseidon, Pluto, and Apollo, and instituted a world order. Similarly, we can imagine their worshippers carefully marking the sacred boundaries between the claims of powerful chieftains. Thus, the living space was divided and life could become orderly.[11]

I understand power (*dynamis*) to be the property of earth who provides fruitfulness and increase lavishly, even with self-devouring profusion. Sky gods render this power effective rather than self-devouring by limiting, measuring, and ordering it. The notion of being, thus, is complex. It includes both power and limit; both energy and a directional or intentional factor which renders the power effective in definite ways. Thus, the historical myth blends with the cosmogonic one. And thus the cosmos, whether within the individual or in the world at large, is given its first and basic shape or direction by an initial vision and selection of possibilities, unseen possibilities being left in the shadow. All else follows from or in harmony with this primary selection.

This term "being' as so far developed is, I am sure, not precise in the way desired by those who are accustomed to think in the language of realism versus idealism. These philosophers will want to know whether this primary ontological notion refers to a matter-like stuff or to a mind-like basis. For myself, I could never see the distinction between these two poles in the sharpened form made popular by Descartes. For whichever of these two meanings the metaphysician selects as the primary one, he will be profoundly embarrassed (as Descartes himself was) by the phenomena of the human body. The difficulty of making the distinction between one's self and one's body (the "lived body,") should suggest that neither can in fact be understood apart from the other; each can enjoy no more than an abstract or perhaps imaginary existence without the other. Thus, it is not the case that the power attributed to being is some sort of physical energy devoid of all direction, whereas directedness is to be associated solely with human mentality. On the contrary, everything to which being may be attributed is at the minimum a directed power. Being is vectorial. Furthermore, the doctrine of evolution can most convincingly be interpreted, not as an argument that the person is "nothing but" material particles in a complex organic structure, but that the elements and species from which man has evolved possess in diminishing degrees and in less explicit form the human powers of intention, feeling, life of which a man is aware in himself.[12] Being understood as directed power is beautifully exemplified in this evolutionary development. To attempt, however, to make the "definition" of being more precise risks committing a category mistake or perpetrating a viscious abstraction. To understand being, we shall best be employed in considering its effects.

Suppose, then, someone were to ask what this effective power effects, could anything be found in the simple universe so far presented which would answer the question? If being is as all sufficing as tradition asserts it to be, then one would expect it to provide such an answer. And in fact it does. Being is the power to effect precisely itself. Being is that which tends to come to be present as just itself. It is self-manifesting. (Indeed, what else is there for it to manifest?)

Being can manifest itself in many ways. For instance, it comes to be present as events and objects of many sorts. But in this essay, with its orientation upon man, his world, and his culture, our concern will be directed upon being as affecting and being affected by and in man himself. Man, as the myth said, is the being who separates power itself from its specific directionality, its effectiveness. By this separation in thought (*logos*) of power from its direction, a change in the use of the power can be entertained. A man can stand apart, as it were, from his own possibilities, view them, and determine which of them is to be brought to fruition. By this separation a man creates "space" for his own activity and achieves a certain freedom in directing and ordering the uses of this power. The nature of this separation of power from its directedness, accomplished by thought and by the instrument of thought, language, merits a place at the focus of philosophic reflection, for by this primordial act of separation human subjectivity comes into existence, man enters a human world, and he takes on the obligation of self-direction. He accepts the tasks of using his powers fully to discover and to manifest the being which he is whether in his mode of self-awareness, in his character, in his thought, or in his action. It also lies within his power to manifest the being of things other than himself.

Perhaps being itself, certainly human being only *tends* toward self-manifestation but may fail in the movement toward this end. Actually to attain this end the power, as one component of being, and the guiding and limiting effectiveness, as the other component, must be perfectly adequate to each other. If they are perfectly adequate, then that which is to be manifested is indeed manifested completely and just as it is. But in the instances where man is concerned these components do not appear to be harmonious. Although one may imagine them to be more or less at one in natural objects, still in the experience of man the nature of the power which he expresses is proverbially a problem, and his efforts to express it are no less proverbially inadequate. Man seems almost by nature to choose his objectives ineptly and to fail in attaining them. Thus, in man's experience the power which is to be expressed seems to be different from (more or less greater than) the effective direction; or the cultural channels along which it is directed are inappropriate to the power they are to render effective. Human motives are ambivalent, and human action tends to become self-destructive. We encounter here the basis of the need for a measure of the appropriateness of action, whether the action occurs within some art or productive function or with respect to the general conduct of one's life. It is anything but clear, as history demonstrates, how this need is to be satisfied, for man's being is not subject to an easy definition which can be used over again as a standard to measure the appropriateness of his chosen life-ways. Hamlet, in this respect typical of Western man, assures us that he cannot be played upon as if he were a pipe (III, ii). We shall find this difficulty to be a major problem.

## B) PERCEPTION

Man's position in the cosmos may be expressed in a story. According to Greek myth there were, among the early offspring of Earth and Sky, a number of intermediary beings, daemons or god-men (e.g., Eros, Hermes) whose function was to carry messages between the gods and earth beings. Now man, in his function of seeing and manifesting being, possesses *logos,* the power of interpretation. He is of the earth, yet he beholds the heavenly bodies and reasons about their eternally regular movements. Thus, he is quite like an intermediary being himself. He is like a minor message-bearer — although a mortal and imperfect message-bearer, — and by means of his interpretative powers he assists (or, perhaps, hinders) the gods in bringing order and completeness to earth.

That being is displayed in man and through him, both actively and passively, appears to be obvious. For man is a being who both affects and is affected. He is passively receptive of that which is presented to him. Also he actively reflects upon, sees or interprets (to himself and to others) that which is presented. In this latter activity, he takes a hand in being's self-manifestation.

To exhibit being, whether one's own or that of another, is to act. When a man acts, he acts upon something, and he produces something. A primary way of his acting is in perceiving. To understand perception, a first step is to discern what the perceiver acts upon and what he effects. Here is not the place to develop a theory of perception, nor to point out the weaknesses of opposing theories, but only to indicate the meaning with which the term 'perception' will be used. Perceiving, I shall say, is interpreting that which one has passively received.

One passively receives presentations. Custom has it that we are receptive only through the five senses. In fact, however, presentations are given in other ways as well. We are aware of such responses as empathetic and synesthetic reactions, as well as of disturbances of the sense of balance, of sensations of heat and cold, of visceral sensations, and the like. In such ways we are given presentations, the *that* which is to be perceived. Anything passively received becomes grist for the process of interpretation.

The presented is the given; interpretation is the active going out to meet this given or the aspect of it to which one has learned to be sensitive. This interpretative process determines in respect to a given world, to a culture, and to the perceiver's intention, *what* that is which is presented. This process takes place upon two levels or in two stages, the involuntary and the reflective. Involuntary interpretation proceeds along automatic and habitual routes to indicate what the presented is. It embraces ways of locating the presented in space and time which are ingrained in the nervous system, as well as ways of identifying objects, their use, and their meaning which are dictated by cultural patterns of response and action and which in virtue of long use have become second nature. I tend, for example, spontaneously to identify this presentation as a pen, useful for writing. No doubt interpretation at this level may be thought to be guided by principles, but these interpretative principles remain implicit upon ordinary occasions of their use. Reflective interpretation, on the other hand, is consciously deliberative. It makes us call upon such more or less explicit principles as have already been acquired or can be discovered or invented for the occasion, and which are thoughtfully and perhaps critically applied to the involuntarily interpreted presentation. Usually the unfamiliar or the problematical call into activity this higher level of interpretation. If the pen is unexpectedly attached to the table, for example, I will be inclined to wonder whether it was really a pen, as I had initially and automatically supposed, and will try to use other evidence to correct the initial perception.

My culture and early training not only have supplied me with immediately available principles for interpreting presentations, but also with habits of rendering explicit, of reflectively judging, and even — gradually and within limits — of correcting my use of these principles. Thus, interpretation is in no sense an arbitrary imposition of meanings upon a characterless given. What the character of this given is, however, can be determined only by testing the initial interpretation against other perceptual interpretations, which in their turn have the *imprimatur* of survival in competition or cooperation with other interpreted possibilities. This survival power has, within my culture, legitimated the interpretative principles which I use and the interpretations of presentations which I am accustomed to make.[13] I trust sometimes hesitantly the experience to which they have led me and that to which they now lead.

## C) EXPERIENCE

For a man, perception may also be experience. If the term "experience" be used broadly, it would have to be understood in its etymological sense; 'experience' is related to the Latin *periri,* and *peritus,* which refer to making a trial, a trial which may possibly not turn out fortunately. "Peril" is derived from the same source. Here experience is experiment, a questioning trial. It asks, so to speak, "What is it that was presented? How am I involved?" The first spontaneous and habitual response to the presented is like a first approximation to an answer to the question: What is the sudden bright something over there? Later interpretations provide more specific and critical answers: "It is a flash of lightning, a discharge of static electricity; at this distance, no danger to me." And I may glance around for further reasurring or warning evidence and for further clues to my own involvement. In short, experience is a more inclusive notion than perception; it is a trial making manifest the object or oneself, usually both. Experience, thus, is not merely perception; it includes self-perception. Through it self-criticism and self-discovery become possible.

A distinction within experience will have to be made between that which is common and that which is special. Neither is easy to define. Common experience is often said to be that which anyone may acquire and is, thus, distinct from the idiosyncratic and from the experience which demands special training and can, therefore, be had only by the specialist. There are many difficulties with the distinction expressed in these terms. For instance, hardly anything is more common than being a specialist. Likewise, that which is common to all at one epoch and within one culture may not be at all common under different dispensations. For his own time, Plato made the distinction rather more clearly. Protagoras in the dialogue by his name tells a myth about mankind's acquisition of virtue (320C-322D). I shall understand virtue to be habitual excellence (*arete*) in the performance of a function. Thus, the virtues are habits; they constitute character and form the channels through which experience is acquired. According to Protagoras' telling, the special arts, agriculture, weaving, medicine, were the gifts of Prometheus, each given to some persons but never to all. With these special arts or virtues alone, however, mankind could not live. Though he could satisfy his individual needs, he could not govern himself in harmony with others. Probably he could see no reason for doing so. To remedy this deficiency, Zeus sent Hermes to him bearing the moral virtues, reverence and justice. These gifts were shared by all, and by their means self-rule became possible. Then in the *Republic* (Books III and IV) the moral virtues, wisdom, courage, and temperance, both as found in the state and in the individual, are further analyzed and unified under justice. Furthermore, the structure of virtue as thus determined is to be found again within any complete specialist art, for each has a productive, a defending, and a legislative function to perform, and these functions, even when performed with habitual excellence, have still to be related harmoniously in one just whole. [14] Here the relation of special virtue to common virtue is the relation of part to whole. Each whole art (with its special virtues) must be related to other arts, and the arts finally must be reflected upon and integrated into the whole of human life and its needs, for the whole of the arts subserves the common good. This final relating function is determined by the moral or common virtues. They determine how, when, for whom, and for what end a special art, e.g., the medical art, is to be exercised.

Now the relation which I have in mind is like Plato's distinction between moral virtue, and the habits required by a specialized art like medicine. Not everyone is expected to exercise the medical art with skill, but everyone is expected to exhibit justice, courage, and the other forms of common or moral excellence in no matter what he may do.

My preoccupation, however, extends beyond the relation of the artisan to the citizen and beyond the four moral habits which insure the good use of the specialized arts or skills. For I am interested in common experience at a level lower than that acquired by Republican citizens in consequence of their possession of civil and moral habits. Are there no experiences had in common at this level of experience? There is the tradition of the *sensus communis* which might be thought to operate to accumulate such experiences. This sense is *le bon sens* which Descartes believed to be equally shared by all. The Eighteenth Century Pietist, Oetinger, following Shaftsbury, distinguished a receptivity to the common truths that are familiar to all men at all times from rational truths of the learned. [15] One might suggest that awareness of being human offers a good example of an experience common to all, but such an awareness is too vague and elusive to help in defining the notion. Common experience may better be approached through consideration of commonly experienced things and events, such things as perceived objects, lived time, existing amidst nature and with others, being aware of obligations, submitting to and reflecting upon the changes which life brings, and the like. Other events commonly undergone are birth, living, and dying. Another excellent illustration is the enjoyment of rite, ceremony, festival, and leisure. For these occupations come into play after the daily (and specialized) work of getting and guarding the necessities and luxuries of life is done. They can make us call upon a people's poetic and dramatic powers. The close relation of the theater to the common life of a people and to its expression in myth and ritual accounts for the repeated use to be made of dramas in the pages to follow. The theater expresses a people's reflection upon its own change and identity. Likewise, rite, ceremony, and mimetic action can operate to bring the participants into touch with the larger world which enlivens and gives value to the arduous and otherwise meaningless work-a-day labor. They cultivate in us a reverence for, perhaps an insight into, the common truths. Common experiences of these kinds will provide a primary source of the data with which we shall deal.

## D) WORLD AND CULTURE

The term "world" can refer to a very large collection of objects or to the scene of mankind's life and activity. Both of these senses make us call upon the ontological meaning of world which renders these other senses possible. This ontological meaning can be approached if it is developed in relation to culture and to experience. Common experience is not of a piece throughout history. We know that it varies from culture to culture. One's culture is an essential factor determining the kind of experience one has, and world is an essential factor determining culture. Culture and world, thus, place an initial limitation and direction upon possible experience.

Since the ways of perceiving and characterizing experiences of men vary from epoch to epoch, then either the being which they are to manifest, or the mode of manifestation, or both, may change. A classic instance of this change occurs in the *Sophist* (274-A), where Plato remarks that that which is, is that

which can be either present or absent. Again, this remark
occurs in the context of the battle of the Gods and the Giants
about the nature of being. The Giants are intractable; they
assert that only that has being which they can grasp with their
hands or see with their eyes. In order to be able to come to grips
with the materialistic Giants, these Giants will have to be
"improved" to the point of understanding that the *way* in which
they grasp and see objects can be either good or bad. And this
goodness or badness makes all the difference. This way of
grasping and seeing, moreover, is not another thing which is
grasped or seen; still, it is a being and involves being just as
much as the physical objects which are grasped and squeezed
with the hands.

The Giants somehow succeed in seeing this point and hence
are radically improved. Before this improvement, the Giants in
their relation to certain powers or qualities of being were rather
like a dog in relation to mathematical entities. A dog may trot
around circles and across diameters, but the ratio of the
diameter to the circumference of a circle (Pi) is never either
present or absent to him. He is not ignorant of it; it simply is
not on his horizon. He has no access to it. The "improvement"
of the Giants must be something like an imagined radical
change in the dog whereby he would acquire the power to live in
a new world, one in which he could envisage such an object as
the ratio Pi. As I interpret the *Sophist,* the improvement of the
Giants marks a change in the basic structure of their subjectivity
in consequence of which they make entrance into a new world
and begin a new epoch of their history. [16] They come for the first
time to see that such intelligible objects as virtues and ideas have
being, can be present or absent, and have the power to affect
other beings and to be affected themselves.

Under what condition can a something be either present or
absent, that is, be able to exercise power upon another being?
In another dialogue, Plato suggests an answer to this question
by reference to place or space (*Tim.* 52B). To be, a something
must be somewhere. But the containing "receptacle" or "place"
(*hypodoke* or *chora*), wherein all beings (including space) can

come to exist, is an extremely difficult notion to compass. Plato speaks of it only in metaphor as a "bastard" notion which we grasp as in a dream and with an effort of imagination (*Tim.* 50A-51B). We can see at the least that this world place where beings appear is not geometrical or physical space. Perhaps we may think of it as the concrete horizon through which ourselves and other things enter our experience or initially become accessible to us. It is the first evidence of a separation between Earth and Sky of the myth mentioned earlier.

Is the emergence of a being upon our horizon or into our world something like ordinary learning, e.g., like a checker player learning to play chess? Or is it a more gradual and difficult process, more like learning to become a participant in a strange culture? Or is it a unique operation, a function of being human and a presupposition of all culture? Surely this latter. Somewhere in the regressive movement back to that which makes experience of a certain pervasive kind possible we encounter that which enables us to develop a culture in the first place. The name I shall give to that which renders a culture possible is "world." World in this ontological sense, thus, is a pervasive structure of human being and of its subjectivity which can be exemplified in a number of cultures. From the point of view of the self, world may be described as the primary relation which a man forms between himself, others, and the environment. Such a relation is not a familiar connection among elements already identified; rather, it is an originative relation, initiating identity and connection. It is basic to any other relation which a man may establish. It is felt and expressed in the symbols of a culture. At times it seems to undergo change.

Culture is a particular and concrete expression of this basic relation called world. A people envisage themselves and their fate in the culture which they have made for themselves. For example, "Western World," as I shall use this term, has been exemplified since the Middle Ages in European and Europeanized cultures. [17] A man can learn and can acquire a culture because he is already the sort of being who has parted the primal parents, so to speak, in a distinctive way and has opened for himself a space for his action wherein he may develop a certain kind of culture. Again, and in terms of the myth, an individual man is in a world just because he possesses within himself the analogues of the primal parents, Ouranos and Gaea, can hold them separate, and can achieve a limited autonomy by use of the culture which he has inherited.

Being, then, is always an effective and limited power. World is the name we give to the most general and pervasive limits which are evident to us. For purposes of this essay the matter may be expressed thus: being manifests itself to man through perception; in particular, man experiences his own being as perceiving. Being, however, cannot manifest itself in human experience all at once. Like the self-destructive fertility of Earth, some of its possibilities cancel out others. Thus, a necessary property of being is the temporality which allows it to be or to manifest itself in partial yet non-self-destructive stages. Being, therefore, is limited and manifests itself only in certain respects or in certain kinds of ways at any one time. World is the most general limitation of which we can be aware which is placed upon being's manifestation to us. A culture is an exemplification of world.

Thus, I shall understand being to be first manifested in a world. A world opens out a limited perspective on beings, some one perspective from among the many which may be possible. In this scheme man occupies a crucial place. He is the one through whom the world's perspective on beings is opened; he is that through whose subjectivity he himself and other beings come to manifest presence. World, thus, is the basic medium through which man is related to himself, to others, and to nature. What I mean is that we do not experience being directly. We experience beings of this or that pervasive sort which our world renders accessible to us. I experience (say) a flash of lightning, not a thunderbolt hurled by an angry Zeus, for I do not live in a world of personalized powers. Being is, in short, experienced through a structured subjectivity. To this basic structure of subjectivity, the name "world structure" may be given. Now, having a world a man can make or acquire a correlative culture and manifest being accordingly.

A world is not a fixed form. It undergoes change; it can become a distorting medium as well as a clarifying one. I give the name "epoch" to a period in history wherein a people continue to share recognizably in the same world. The structure of this medium, then, different in different epochs, determines *what* can get through to the self and how it can be accepted, associated, valued, used, in short interpreted.

The study of the world-structure of an epoch may, at least in some of its aspects, be approached by determining what is common to a group of cultures. What I shall call the Western world begins somewhere in the early Middle Ages, continues into the present, and includes the later Medieval, the Renaissance, and modern cultures and their several sub-varieties. It is characterized by the following common elements of structure: it values men intrinsically, although this value may be variously determined, e.g. theologically, morally, politically, economically; it regards the self as relatively indeterminate in nature and the self's completeness, at any specified time, as uncertain yet achievable by some (variously identified) supra-individual aid; and it understands man to be essentially related to nature, whether as guardian, knower, ruler, or user. In our time there is some evidence that this world-structure may be undergoing radical change. We may be moving from a world in which man dominates his technology to one in which his technology dominates him (cf. §35). If so the matter merits attention.

This view of the world is opposed to the conviction that the world is an external something to which one reacts passively. It denies that the world is a kind of show, a given which is merely observed, theorized about, predicted verifiably, and controlled, but which is independent of these activities. Perhaps this latter view is expressed by B. F. Skinner when he remarks, "No theory changes what it is a theory about."[18] It is not quite clear, however, what this remark means. Surely it is not the simple-minded caution against thinking that the sun will rise will make it do so. For everyone understands that a theory is usually about something other than itself. This other, however, is experienced and known only through the mediation of a world and the accumulated knowledge (interpretative principles) which determine what and how one observes and how one theorizes. The relation of seeing to the thing seen is not the simple relation of reflecting to reflected. Consider views of the sun. Amenhoteph the IVth. held it to be a god; he *saw* it as such. For Plato the sun was the origin of the visible world and the analogue of the Idea of the Good. For a modern man the sun is seen as a minor star about midway through its life span, located at one of the foci of our solar system. Each sun belongs to a different world and each appears foolish, metaphysical, in-consequential, or false when transposed to other worlds. Of the sun itself, one might say that it is such as to appear in the several ways which these different worlds allow.

One may also think of the world as the stage upon which the human drama is played. It is a stage which is intimately linked to the actors themselves and to their play. Although one cannot say that this stage is devised by the players, still it affects the players and is affected by them in complex and often obscure ways. The props on the stage, the geometry of the stage itself, all fit the grand plot and figure in the movement of the action to its end. I do not exclude the possibility that an aspect of some plots may be the supposition that the stage is the main player, the protagonist, and that he *seems* to play to the tune called by his human authors. In fact, we shall be especially interested in seeing how such a play develops.

It will be convenient now to summarize the route this essay is to take by continuing to borrow terms from the theater. Our theme will be the existence and fate of the self upon the stage of the world which we now inhabit. Several chapters (II through V) will be required to treat of this protagonist. These chapters, in keeping with the conviction that action is a basic mode in which human being is manifested, will follow a modified Aristotelian schema. They will distinguish role-player or specialist from the self of common experience. Then they will consider the latter's character, his thought, and finally those aspects of plot in which he cannot but be involved. These discussions will terminate in an account of the self which, without being a definition, will nevertheless suffice to display the conflict between the humanistic and certain technological tendencies of our tradition and for executing the measurement in which we are interested.

A sixth chapter will turn attention upon some of the notions basic to science and technology and will develop several senses of the concept of measurement. Here the concepts of qualitative and of quantitative measurement will recur and will be considered in greater detail than in the sections 2 and 3 above. The final chapter will take technological man and his culture as the antagonist. Technological culture will be defined, and the measurement of that culture and its concept of man by the previously reached understanding of man will be offered. Here will be taken the (qualitative) measure of the culture whose primary effort is directed toward the development and application of the (quantitative) sciences, and the goal of this essay will be reached in so far as it is successful.

## §5. *A Note on Method*

I append a brief remark on the way chosen to the goal of this essay. First I observe that no special attempt will be made to prove or to establish the ontological principles laid down in the preceding section. Rather these ontological principles will be presupposed and utilized as a guide to interpreting and unifying the data and notions to be adduced later. I must add, however, that I am far from supposing that the usefulness of ontological principles in interpreting and unifying human experience is irrelevant to their validity. Indeed, if this applicability is demonstrative, then this whole essay may be taken as a demonstration of these principles.

For the rest, I shall attempt to see that the method used is, following Aristotle's injunction, appropriate to the problem and to the data.[19] The usual techniques of critical analysis will carry us some of the way. But they will not provide the evidence in

favor of the route which I have chosen. In addition, therefore, I
shall employ phenomenological description.

In order to approach that which is to be described,
phenomenologists often employ an epoche or suspension of
certain basic presuppositions. The best known instance is
Edmund Husserl's suspension of the universally presupposed
existence of an external, objective nature. Suspension of non-
use of this presupposition leaves the phenomenologist present
with the given phenomena as they are presented to his consciousness. Husserl thereupon sought to recover a grasp of
constituting subjectivity, to unite this recovered part of his
being with objective cognition, and thus to reconstitute the
whole of human knowledge and experience. He sought to
develop his analysis of the phenomena thus presented into a
transcendental philosophy in which all that had previously been
suspended was recovered in the guise of meanings indubitably
present to consciousness.[20] In this essay, however, my purpose is
not to discover the absolute origins of experience or to elaborate
a transcendental philosophy; although, this transcendental
level is approached by another route in §19. I shall not,
therefore, need to execute the epoche of basic presuppositions.

It will be useful at several points, however, to perform a
partial epoche. By partial epoche I mean one which is not
intended to reach an absolutely primary level relative to the
whole of thought or experience but one which, nevertheless,
exhibits the phenomena in question upon a more basic level
than that prior to the epoche. This epoche will suspend certain
normally used distinctions. For instance, I shall when necessary
suspend the conviction that specialized experience (e.g., the
scientist's experience) is more authoritative than common
experience or that scientific and technological knowledge are
more reliable than common knowledge about common matters.
(By a sort of reverse epoche, one may think of the specialist as
suspending common experience in order to gain access to the
specialized data which concern him; the specialist's suspension,
however, often goes further; it becomes a denial of the validity
of common experience.) This suspension is a device for short-

circuiting specialized experience and knowledge without rejecting or denying them. In consequence, ordinary or common experience should become available for inspection and description. Likewise, I shall suspend the conviction that in all one's activities one is engaged in playing some sort of role. By putting aside this conviction, without denying it, I hope to approach closer to the experience of self. Thus, something like self-experience will become accessible. And in Chapter V, I shall question the conviction that many events ordinarily considered to be necessary and significant are in fact necessary; thus, I shall be able to isolate the absolutely essential events of one's life.

These partial epoches should open up correlative areas of phenomena for description and analysis. By description I refer to an interpretation or translation of the experienced phenomena into language. Experience, as observed above, is already an interpretation. Description, which attempts to be a further and faithful translation of the experience is, thus, a (linguistic) interpretation of an interpretation. It seeks to render the experience, especially at the level of involuntary interpretation, explicit and communicable. Translation involves the important notion of correspondence and will be discussed in section 18; also it elicits the difficult problem of distinguishing the literal from non-literal uses of language and of justifying the latter use (cf. pp. 183-7). Such descriptions as I produce, or reproduce by reference to other writings, should be sufficiently intelligible; they usually are offered as data supporting the proposals, theories, or evaluations which are advanced. By application of this methodology, I want to explore the existence and fate of the self; that is, to examine the power and limits upon the power of the self to interpret itself and to direct its own life. In particular, I desire to determine whether the ways in which life is presently being directed within the Western world are the most profitable among those now accessible to us.

CHAPTER II

# ON THE NATURE
# OF THE SELF

## §6. *Self and Object*

W e who undertake the measure of technological culture have ourselves been formed by that culture. Since childhood we have heard the praises of this culture knowledgeably and persuasively sung. Perforce we are convinced; nothing further can be added on that score. In addition, we have heard it negatively criticized. This culture is alleged to place destructive power in the hands of irresponsible people and their politicians, to hand the environment and the laboring population over to self-interested economic entrepreneurs,[1] to destroy the capacity and the will to work by mechanizing the conditions of work, to eliminate the possibility of personal growth by requiring total devotion to a specialty, to foreshorten the vision of a humanly habitable world by generalizing and reifying the concept of a predictable and controllable nature. How far are these criticisms justified? In particular the last two, which concern man's grasp of himself in a habitable world, are difficult to understand and apply. What is the self's relation to its world. What, indeed, is a self? These two questions broach our topic.

"Enormous repose before, enormous repose after the flash of

31

activity," but who can say what the activity is? or the repose? A guiding declaration which will, at least, keep us to the question is the following: man is the being who is aware of an obligation to understand the truth about himself and about the essential human events. The main terms of this declaration indicate the topics for this and the next three chapters.

The present chapter will be occupied with self-awareness. Also it will be concerned with the context — the world — within which a person experiences himself as he engages in self-interpretation. Any person is subject to being diverted from the self-understanding he seeks by the roles he plays or by the ways in which he plays them, a diversion sometimes rendered all but inevitable by the world he inhabits or by the culture he develops. Perhaps he can guard against this diversion by forming a concept of the self which will guide thought and activity. Unfortunately, though there are reasons — this chapter will present some of them — for believing that a clear and precise concept of the self cannot be reached. The question, then, "what is a self?" cannot be answered in a straight-forward manner. In our world the dialectic of self-interpretation leads to another kind of grasp upon the self and will eventually provide another kind of standard for evaluating a culture.

What, then, is it like to be a self? I believe it to be undeniably true that each of us has some feeling, some perception, however inchoate, of the self which he is, but I find it extraordinarily difficult to indicate, much less plausibly to describe the presentation of the self. Nevertheless, our political, legal, medical, and educational systems seem to hold it obvious that each of us is indeed a self and is able to sense and to interpret himself, at least in an everyday manner. But beyond the safe confines of these conventional beliefs, questions about the nature of the self invite all the confusions which attended the opening of Pandora's box. When I open this box I am immediately confronted by such definitions as these: "man is a rational animal," or "man is the language-using animal," or "man is the omnivorous biped that wears breeches." When I put these definitions aside, thinking perhaps to find the self underneath, I find instead such metaphysical concepts as these: the self is an imitation of a form, an individual substance of a rational nature, an image of God, a thinking thing, a succession of impressions unified by convention, a machine with complex feedback arrangements, a product of work and interacting social and natural forces,[2] a something which learns to use the first personal pronoun correctly, a consciousness free to determine itself but which fears and avoids use of this power. I put all these doctrines by since I suspect them of being prejudices. And I suspect them of being prejudices if for no other reason than that the proponents of each have repeatedly attacked and presumably destroyed every doctrine except their own.[3] Put all together, all are destroyed. Likewise I put aside scientific views of the self which are built up from a careful and systematic selection from the evidence interpreted with reference to highly technical assumptions. By a sort of partial epoche, I desire to separate the self from these limited expressions and specialized interpretations. I would like to hew to the self as it is and has commonly and widely been accessible in Western cultures.

At the outset, however, a difficulty is encountered. The assertion, "I am aware of myself," betrays it. For the "myself" in the objective position seems to be an object of knowledge, something intended, definite, perceptible, present in a familiar context. At the same time, the self, the "I" who is aware, the intending "I" without whom there is none to claim experience of self or of objects, is altogether different from an object. Confusingly, then, the self is present in two unlike guises.

That the self is object to the self and in this capacity bears analogy to other objects is easy to see, for the perception of self initially seems to be analyzable like the perception of an object. Every man, I shall continue to suppose, can have a spontaneous, pre-rational awareness of his persistence in being; likewise he undertakes, after reflection, to engage in some activity, a trade or profession, a mode of life. He senses *that* he is and asks *what* he is or may become. It will be recalled that the perception of an object was analyzed in a similar manner, into spontaneous and reflective interpretations of the presentation (§4B).

Yet in spite of this similarity, the self cannot be merely an object. For in self-awareness the self is present to itself in an altogether non-objective manner. The self, unlike an object, proceeds to its own interpretation. As touching this topic, Descartes' reflections are paradigmatic. He observed that he intuited his own existence, and noted that this intuition was always possibly present and was actually so upon any occasion of his doubting or thinking anything. He then turned quite naturally in the Second Meditation, to interpreting this given, asking *what* this immediately presented *that* might be. His conclusion, that all awareness is essentially mathematical thought, illustrates what is meant by reflective self-interpretation. I do not pause to criticize Descartes' view of the self (this whole essay may be taken as such a criticism); the point is this turn is altogether normal. Anything one might imagine or say about the fact of one's self-awareness constitutes reflective self-interpretation, even if the interpretation were false. Both intuitive awareness of one's existence and its reflective interpretation are non-objective acts, well beyond the range of behavior attributable to objects. They are activities of the intending self; they are not objects; they are not, for example, the objective or intended self.

Still, other aspects of the self are undeniably object-like. The initial intuitive self-presentation eventually develops not only such reflective interpretations as Descartes' "I am a thinking thing," but also such as these: "I am an engineer" or "I am the elder son of my father." Thus, the presented self is interpreted in terms of an elaborated world view, a familiar profession, or a kinship system. Consider the general case: in a social group, a person aware of himself as a part of this whole and perhaps obligated to it, tends to interpret himself by understanding himself in an accepted manner and by assuming a role, a function, or profession which is defined clearly enough within that social context and possesses definite object-like properties. By thus acceding to the demands of role and of the social forms which define it, the elusive self seems to be caught and fixed like an object in the net of a familiar structure. Many other kinds of self-interpretation similarly objectify the self. Perhaps, then, the self is an object, albeit an unusually complex one.

This conclusion may seem to be reinforced by the self's intimate connection with a particular object, its body. For surely one's body is an object. It occupies space, moves around, can be bumped into just as one might bump into a table. It has the properties and life cycle of any organic body. I seem, morever, to be completely identified with my body. When it moves, I move; and as I move, so does it. Where, when, and how I am, thus also is it. To attack or to protect my body is to attack or protect me. This identity, however, is ambiguous. If my finger or any other part of me is cut off, I no longer feel identified with it but only with the larger remainder of my body. The severed part, we say, is unowned, is dead, for I cannot direct its movements. When, in the course of time, my whole body comes organically to resemble my severed finger, then I shall be pronounced dead; then my capacity for self-direction will have been severed from all worldly power. The self, thus, seems to be identical with no (detachable) part of the body, yet it depends for its very life upon the body. What is the nature of this strange relation of self to body? Nowhere have I seen a clear answer to this question, but certain observations are easily made which suggest that this relation is an altogether unique one. The self who intends objects is itself not an object (for objects do not intend); yet it is aware of itself only as embodied in something which is very like an object and a part of nature also. Through the mediation of the body it encounters and communicates with others whom, however, it recognizes to be non-objective.

Whatever the self intends, it intends through its body. It has a world only by means of its body. In sum, my body is the effectiveness of my being. Thus, self and object meet and commingle in the human body. The embodied self is part of nature and in many ways resembles an object, yet only a non-objective being can recognize itself as part of nature or be aware of its resemblance to an object. Thus the self is not easily caught, defined, and objectified. The engineering skills which qualify a person to play the part of an engineer are not given like objects; the engineer gave them to himself by choice and by self-discipline. The Prince of Denmark immortalized the difficult decision to become his father's son quite late in his history. My point is that objects are not presented to themselves, do not feel themselves or think, are not aware of the functions or roles which they play, do not interpret themselves in a kinship system. Thus, in the sense in which the self does all these things, it is not objective. The power of the self to stand away from the ongoing process in which it is involved, whether in vague self-awareness or with a critical effort of self-interpretation and of self-direction, is not merely another, though complex, object-quality. It is a power of a different and non-objective kind. Unlike any object, the self possesses subjectivity; it is only ambiguously identified with a body, its own. And though the self may cultivate an objective attitude or play a role as if it were an object, still it can always reinterpret itself and change attitudes or roles without loss of self-identity.

However closely, then, a person may seem to have related himself to an object-like body or to a routine, trade, or profession, it is always true that he interprets himself in this fashion and that he may possibly interpret himself differently. He is always subject to radical change. In other words, although a man may be possessed of object-like qualities, he also has a non-objective dimension, and hence his self-interpretation, both at the immediate and reflective levels, develops differently from the only remotely analogous interpretation of object presentations. Self is intending of self. Objects are interpreted; selves are self-interpreting.[4]

No interpretation, however, whether of objects or of self proceeds *in vacuuo*. It proceeds in a context and within a culture whose general and characteristic properties are determined by the environing world. As a man stands off from himself in self-awareness or self-criticism, he is not dissevered from himself. At the minimum he remains and must remain in his world. The world, thus, relates the self to the self. Let us, now, take notice of the ways the modern world inclines the self to interpret itself.

## §7. *Self and Worlds*

A. N. Whitehead remarked once in a lecture that the single great metaphysical notion which originated in the West in post-classical times is the notion of self. The observation is reasonable, for it is likely that self as it has been experienced in comparatively recent centuries could not have been present in the Classical Greek or earlier worlds. Both Greek and Christian sensed that the individual self is initially a failure, a non-being, but that it is directed toward manifesting the being which it is not yet. Beyond this schematic similarity the difference between Greek and Christian is a difference of worlds. To the Greek, man's initial non-being is a quite definite potentiality, a potentiality, moreover, which a man, exercising the appropriate rational discipline, could actualize himself and thus achieve that harmony of soul which is the perfection of humanity. This is a doctrine of self-salvation. Quite otherwise with the Christian. The Christian is plagued by the paradox of man's finite power set in contrast to his infinite destiny. He senses his existing self as indefinite, a loss, an absence. Man's nature is hidden from him; he is only a partial being. His completeness is vanishingly remote, not to be attained except through help from without. This Christian doctrine, therefore, must be a faith in salvation by the other. If man is self-interpreting, Western man must add that his self-interpretation is intrinsically dependent upon another, both for an adequate grasp of its direction of development and for power to attain this end.

Not the Greek movement to perfection but the Christian longing for an indefinitely remote completeness was destined to be inherited by the Western world. Although Plato may have reached some apprehension of what was later felt as self, it is probable that Socrates did not. Socrates, gad-fly of the state and ironist of the market-place, was said to be like the images of the Silenoi sold in the streets of Athens (*Sym.* 216D). Like a Silenus, he was rather repulsive outwardly, yet within he contained golden images of the gods. These images, of course, stand for the virtues, the principles of orderly life, which perfected the citizen of the polis. Socrates was a model at the banquets of men, but only seldom does he express a sense of absence from the banquet of the gods (but cf. *Phaedr.* 247). On the whole, as Mathew Arnold remarked, Socrates is terribly at ease in Zion.

Or is it not more likely that Socrates was never in Zion? That later Platonist, Saint Augustine, however, understood what it was to be in Zion, even though he was in Rome or Athens. By the time of Saint Augustine, the new dimension becomes evident in the world of Western man. Saint Augustine, though trained in Greek philosophy and rhetoric, apparently succeeded in translating every item of his life into the Christian idiom. Robbing a neighbor's pear tree came to be a world-symbol of the Fall of man and of his remoteness from God. [5]

This remoteness is intensely felt as an emptiness of his own making; thus his being is perceived as defective to the very core. Yet, as a Christian, he also believed that the way to filling this emptiness was open to him and was indicated at every turn. For in everything an infinite possibility was envisaged; everything in the world, to the understanding soul, speaks of the being on which it depended. Merely, for example, remarking "I am," could be understood to indicate the guiding presence of the Holy Trinity within. [6] For one of Saint Augustine's faith, not only the self but everything, every event, becomes fraught with a meaning unknown to the Pagan world; every moving thing refers to the infinite Wisdom and energy by which it was created. Thus, every object in the world calls to the erring person and awakens a longing for completeness of being and for reunion with his source and his end. [7] But this source and this end are infinite. What is man in relation to this infinite being? Man should comprehend, Pascal said, that he is in-comprehensible. And Saint Augustine had already remarked, *"Quaestio mihi factus sum*—I have become a question to myself." In any event, a person is not neatly definable by an essence into which he will certainly grow if only he conducts his life skillfully and morally. Modern man has left this Greek world behind for a different one. It is as though this progeny of Ouranos and Gaea (§4A) had, in forming their living space, separated their parents so far that no relation between them could be discerned by the unaided eye. Impotent man now has to wait upon the initiative of some other power if he is to see the way to his own completion.

The new dimension which the Christian world opened within the self might be described as a vivid sense of the depth and mystery of human existence and fate. It is exhibited not as a sense of possible perfection nor as a drive to excellence. Such a drive, which we associate with the Classic Greek, depends upon a conviction of the adequacy of the present and available world. This consciousness the Greeks, with their belief in the holiness of earth and in the sufficiency of the order emanating from the immortals of Olympus, undoubtedly possessed. But the Christians added a sense of time. The present world is incomplete; it is in need of something not now present but possibly to come. Existing man, too, is radically incomplete; he is alienated from the whole of his being and awaits the gift of a new life. He is a pilgrim upon this planet. The Greek ideal of finite perfection is replaced in the Christian world by the ideal of an infinitely remote completeness. The nisus of post-Classic man is for a wholeness incommensurable with anything within reach of ordinary experience.

Often it is difficult to know, when examining human development, when one is faced with an actual change of worlds or when the change is only cultural. One important criterion involves the self: if the change leads to a radically new way of understanding the self and of using the human powers, then it is a change in world structure. If, though, the change is an alteration in the mode of using powers which are already familiar, then the change is cultural. Thus, for example, the materialistic Giants of the *Sophist* (cf. §4) underwent a transition into a new world when they learned to perceive a new kind of being, Ideal being. At the same time their character changed; they became able to be friendly with the Friends of the Forms, and doubtless they envisaged a new significance in all their actions. As it appears to me, the movement from classic Greek to Hebraic-Christian marks just such a shift in worlds, for here the contrast is between two kinds of being which man attributes to himself; the shift is between a man who can perfect himself through thoughtful practice, who can bring himself into harmony with his fate and become divine, and on the other hand a man who finds a measureless depth, an emptiness within himself, and who recognizes a need for help from beyond himself, from his source in a greater being, in order to be himself. Here we encounter, I suggest, two different worlds which render possible quite different kinds of cultures. I find it impossible to imagine a technological culture developing in a Greek world. But as history has demonstrated, a redirection of certain basic drives was sufficient to open the way to a technological culture in the Post-classic or Western world.

Now the inevitable consequence of the thousand years schooling in Mediaeval Christianity is that the self acquired and still owns to awareness of an inward transcendence toward wholeness. A claim to this drive remains a part of modern common experience, and self-interpretation is inevitably moulded by it, even up to the present time. This drive, however, is no longer always expressed in religious terms, but it is nonetheless a present and determining factor in the structure of the modern world.[8] It is expressed in the Gothic cathedral, with its movement upwards to the religious infinite. It was present in Renaissance sonnet sequences addressed to unimaginably pure and unattainable women and in the infinite mathematical universe which was opened before man's intellect and his imagination. This drive scarcely grew fainter with the ambition of the Enlightenment to reduce the whole universe to a scientifically intelligible system. Likewise, it is present in Goethe's *Faust* with its ceaseless secular striving. And surely it continued an active life among the Romantics, with their zest for infinite tasks and their resolution to save the depths. Also it was mirrored in the Romantic vision of a richly creative nature, endowing man's home with wonders and rivaling his creative afflatus with its own endless increase.

There are those who find in the contemporary world something new on the scene, a mechanical world, product of the sciences, instrument of the technologies. But that it is something totally novel, not sharing in the nisus for completeness which came into history with Christianity, is not evident. For this mechanical, this empirically necessary and impersonal world, so ideally suited for use in simplifying and resolving problems in the sciences and technology, expresses the same nisus for completeness, only in a displaced form. It is redirected from man himself to the notion of complete control over an external nature. No longer is man himself to be changed and so to be made one. But a raw nature is to be changed and brought completely under human control. Man sees, subsequent to this change, his own completeness in his dominion over nature. This alteration in the way the self views its relation to the world, while not yet total, is still no trivial change. I suggest it may approach a basic change, one which may well inaugurate a new world and a new epoch in human history (cf. §35).

Whatever his future, currently existing man exhibits a secularized version of the Mediaeval world-structure. What can be said of the nature of such a self? By what process does it interpret and develop itself?

An initial and very general clue to relevant kinds of truth about the self is offered by the tradition which holds that man is always the living being possessing *logos*. [9] Whatever world he may live in, he is the power of interpreting, whether he be speaking or silent. Indeed, the possession of *logos* may be essential to the power of entering a world and forming a culture, whether the world be Greek, Christian, technological, or any other. Let us, then, consider what the self must be like if it is self-interpreting in a manner consistent with Western-world cultures.

§8. *The Self as Self-Interpreting.*

An object is interpreted; a self is self-interpreting. Still, the conviction persists that a self interprets itself as like to an object, even though, perhaps, significantly different from it. [10] How great is this difference? Herein lies an essential question. Let us for the moment accept the analogy of the self to an object while keeping the common experience of humanity carefully in mind. An experienced object is a presentation which is spontaneously, then reflectively, interpreted (§4B and C). To retain the analogy, we need first to identify the presentation of self and then to examine the ways in which its interpretation develops.

The presented element in object-experience is the given. I suggest that this given element in the experience of self is simply the self's possibilities.[11] More particularly, this given is the self's possibilities of manifesting its own being. These non-objective possibilities or powers are, as in the presentation of an object, not experienced as such. Inevitably, some particular interpretation is present, however inarticulate and a-rational at its lowest level this initial interpretation may be. The question is wherein this initial interpretation of one's indefinite possibilities is to be recognized. My suggestion is that the spontaneous interpretations of self is to be identified within the flow of feeling which we all constantly experience (cf. pp. 132-4).

I take the flow of feeling to be the first aspect of experience which betrays awareness of self. The feeling is sensed as localized; one might say, as "mine." Likewise, it is an aspect of experience which, except in sleep, is inevitably present. Feeling, the intuitive sense of one's possibilities, the awareness of "power to—" or "power not to—," is the first response to that to which the self is sensitive and thus offers a kind of index of the current self. And though this feeling response is non-reflective and does not answer the question concerning what aroused the feeling, nor what that is in which it is aroused, still it may elicit these questions and serve to invite or to block reflective interpretation. I say "my pen" and speak the phrase with some feeling, however slightly the feeling may have been noticed. I speak the phrase with anxiety, with pride of ownership, with surprise, with irony, or the like. Such feelings constitute part of the non-objective reference of the phrase. These feelings refer tacitly to the self, to its own way of making the objective reference to this object, my pen. They identify me as the self who makes just that sort of reference to myself as well as to the pen.

Although feeling is implicit and indeterminate, compared with a precise concept, it is not altogether vague. In all feeling, in addition to its tone and particular reference, a sort of general intentionality is detectable. For instance, all feeling, over and above the property in consequence of which it is immediately recognized as guilty, self-approving, boring, content, stressful, and the like, expresses one's intention to manifest one's own being. All of one's impulses and actions, however particular, can be observed to express this general and driving intention. Such expression is necessarily limited to the present time, its opportunities, and its accessible possibilities. I shall refer to it as the existing or current self. A second generalized intention is also detectable within the flow of feeling; this intention is in our world the nisus for wholeness of being; it is the intention of completeness. These feelings are one's initial self-affirmation and express the persistent effort of the current self to become a whole self, whatever transformations may be required. Commonly and primarily one seeks to discover, to preserve, and to bring to fruition one's own characteristic way of being, of feeling, of seeing oneself. This fundamental sense of self, of its unity, and of its need for completeness, runs, except in abnormal instances, like an Ariadne's thread through all experience. Of course, these fundamental intentions can be blocked, not merely by natural catastrophes, but more especially by hostilely disposed men, by unintegrated elements within the self, as well as by frustrating elements within cultures. In response, these original intentions can be mobilized and provoke, no doubt, the power behind the many social upheavals which are regularly described as striking a blow for the freedom of men to be and to determine themselves.

This conviction, that one first comes upon the self in the flow of feeling, is confirmed by the opinion of some psychologists. For example some psychiatrists make the assumption that their patients possess a latent yet sufficiently definite apprehension of their identities to reject a hypothesis about themselves or their

motivation which is clearly wrong or out of character. Hypnosis
also provides evidence to this point. If a person under hypnosis
is told to act in such a way as to deny a strongly held feeling of
his identity, e.g. if he is told to do something which violates a
firm moral conviction, he will awaken rather than carry out the
command. This inarticulate awareness of self, according to the
views of some writers, also receives an initial and common
expression in the symbols of dreams and of art[12] as well as in the
symbols belonging to one's world (§27). I am suggesting that the
spontaneous awareness in feeling of one's possibilities and of
one's completeness, likewise the expression of this immediate
awareness in ritual, myth, and symbols, may offer a more effective guide to one's wholeness and maturity than many highly
wrought and successful theories (one such theory is outlined in
§36). In any event, although one's flow of feeling is relatively
indeterminate, still it is sufficiently specific to follow an "inner
logic." It is a readiness for certain acts or for a kind of experience; it is a sense of "power to—" or "impotence to —"
which expresses one's intention to persist and grow in one's own
being. And clearly not every imaginable manner of being
oneself is consistent with this initial sense of one's possibilities.
Choices must be made. It is essential that awareness of the sense
of self be sharpened so that it may be used as a guide in making
choices. For otherwise biased theories, group pressure, family
preferences, emotionally charged fantasies, and the like will
determine choices. Then that which will be manifested will not
be one's own being but that of the other agency which was
accepted and substituted for the sense of one's own wholeness.

My hypothesis, then, is that the self is an activity or directed
power, whose function is to manifest its own being. Initially the
self is its possibilities of self-manifestation; these given
possibilities first emerge in feeling, for instance in the sense of
"power to—" or "impotence to—." And further, this feeling for
one's possibilities is given its basic expression by one's world and
world-symbols. It is manifested to us as the obligation to pursue
the concrete truth of oneself. One must strive to know this truth
in order to attain one's good.

This humanistic hypothesis is useful not so much because it solves problems but because it provokes questions. Which of the self's possibilities are those which primarily manifest its being? In what direction will concrete wholeness most reliably be sought? What is the self's relation to those possibilities of action which do not primarily envisage its own being but which nevertheless belong to the self and can preoccupy its attention? How do self and self-interpretation develop? How may this non-objective dialectic go astray?

Precisely because the answers to these questions are not laid out unambiguously in a definition or in a picture of the ideal man, we considered that a man is well advised to regard himself as obligated to pursue the truth about himself. In order to deal with these questions so far as needed to develop the measure of a culture while retaining our ideal of concreteness, we shall have to ask how the person experiences himself or others at a level higher than the inchoate self-awareness present in feeling. And we shall have to remember that, within a finite lifetime, many of one's possibilities will remain merely possible, and thus the experience of the self will always be deficient. Also it may be that the more significant experiences of self will remain among these unrealized possibilities. [13] Finally, owing to the limitations imposed by our world view, awareness of some possibilities is normally excluded from the beginning.

## §9. *The Dialectical Nature of Self-Interpretation*

Worth emphasizing is the fact that all of a person's possibilities cannot be expressed at once. We live in time, and some possibilities must be made explicit and cultivated before others become accessible. We live in a world which admits some possibilities and denies or ignores others. From out of the flow of feeling, which is at once the individual's initial expression of his possibilities and his response to events, some feelings are selected to receive further and reflective interpretation. But it is to be noted that this reflective interpretation, whether it be expressed in action, imagination, or critical thought, provokes a still further basic response in feeling. That is, we respond with new feelings to our own reflective interpretations. These new spontaneous responses may lead to further reflective interpretations. And thus there develops a dialectic between spontaneous and reflective interpretation.[14] What child has not experienced something like a curiosity about (say) clocks, a curiosity leading to disassembling and finally to a successful reassembling of a clock. The excitement thus provoked might well lead him to further feats of mechanical imagination and experimentation. This particular illustration of dialectic between spontaneous and reflective interpretation emphasizes the objective; it is outwardly directed and is often focused upon and powerfully reinforced by its obvious relation to ends which are publicly esteemed. But a similar dialectical development also leads the feelings most closely related to the self into expression of a self-image or self-concept. This self-image and the practices and roles through which it is rendered concrete elicit further feelings whose self-reference may be developed into further self-interpretation. By this non-objective dialectic, the dialogue of the self with the self, one's grasp upon oneself and one's world grows. But this kind of growth is subtle; it is not open to easy inspection and seldom receives public rewards. This dialectic, or inner dialogue, nevertheless, is the road to self-discovery; it is the way by which the existing self approaches the complete self from which it is alienated.

The matter may be put thus. Early man observed himself hunting and gathering food, compared his techniques with his neighbor's, and improved his own. Perhaps he even learned to hunt his neighbor and to gather his neighbor's food. And so the objective dialectic developed. But the matter progressed a step further. The same man stood, so to speak, outside this movement, observed himself observing, and asked: "Who is this being who both acts and critically observes himself acting? How can he both be himself acting and outside of himself observing? Does he not thus become another self?" And so the existing person finds himself engaged in a non-objective dialectic of self-change.

Growth of this kind in grasp upon onself may accompany the more visible and familiar objective dialectic by which one learns a role or acquires a profession. Unfortunately, though, it may not do so. The non-objective dialectic may be misdirected, it may become atrophied, or it may become dominated and distorted by the objective, the consequence being that reflective interpretation of the self fails to express, reinforce, and develop the immediate experiences of the self's possibilities which are first registered in feeling. One's bond with the world and one's vision of a possible whole self become enfeebled. The literatures of Western countries are rich in illustrations of this failure. Arthur Miller's "Death of a Salesman" and Herman Hesse's *Steppenwolf* provide examples. According to some analyses, this exiguous self-development is endemic in the West.

Whatever the fortunes of these two kinds of dialectic, the objective and the non-objective, and however happy or disharmonious the relation between them, they combine to form what I have called the existing or current self. Thus the phrase "the current self" refers to that selection among one's possibilities which has emerged as a felt intention to be oneself and which has received reflective interpretation as the set of roles which one is playing at a given time.

The question how one becomes this current or existing self rather than that is always pressing. Somehow one set of possibilities is selected for dialectical development and others are rejected. By what means is the selection made? We have noted that one's world allows only a limited set of possibilities. The culture in which one is bred produces a further limitation. From within the remaining options, varying means operate to select the ones which seem eligible to the individual person. One way of selecting out of this set is by use of techniques of generating a vivid experience of the self which one is not yet but which one might well desire, or be brought to desire, to become. I refer to poetry, drama, and in general to rhetorical techniques for inducing by linguistic and mimetic means an initial grasp upon possibilities as yet implicit and unpossessed but inherent within the self.[15] Thereupon, these vaguely apprehended possibilities of self are given concrete expression and tested in the theater of daily life. Folk tales of a hero encountering and surmounting dangers, acquiring virtues and magic powers, offer a mythic version of this movement toward self-completeness (cf. §22). These rhetorical techniques are to the humanistic tradition what operant conditioning is to a more recent school of thought about the self.

The speaker or writer using humanistic techniques may perform his communication by use of metaphor and by advancing analogies to experiences which the listener has already had. An artist, for example, being one skilled in communication without use of abstract concepts, may express his relation to his art to a successfully married couple by saying that he is married to his art. But if the couple in question were and had always been merely role-playing in their marriage, then they would have no corresponding feelings about themselves and their marriage to provide the basis for understanding such an analogy. Hence a dialogical communication with them, concerning the artist's relation to his art, can on this footing scarcely be begun. And similarly with respect to speech concerning persons, Antigone could not communicate to Creon her personal obligation to her slain brother, doubtless because Creon was accustomed to dealing with people exclusively in their roles as subjects or as instruments of his power. And unfortunately Antigone was no Sophocles. She lacked the art of awakening those possibilities in Creon which would have opened the way to an understanding between them.

Perhaps, therefore, it might appear that rhetorical and literary devices can communicate only about that which has already been experienced. If this limitation obtained in truth, then one would be denied this mode of access to those possibilities of one's own or of others which had not already been enlivened. But in fact, experience need not be a mere repetition of the previously acquired. And in fact, the modes of communication just mentioned do impart insight into one's latent possibilities and can release their power. Herein lies the force of rhetoric, the theater, poetry: these techniques can communicate new experience; they can open up new possibilities and elicit new kinds of feeling, new self-interpretations. Indeed, they sometimes play precisely upon the possibility of radical self-change; this is the change which must be recognized to be an alteration of one's total response to the total environment.

The reader will catch my meaning if he recalls the dramatic events by which King Lear moved to an awareness of Cordelia's existence as something more than a charming puppet playing the role of dutiful daughter; at the same time Lear himself moved to a more complete grasp of himself as something quite other than an elderly Don Juan. There would seem to be a good probability that a reader of this drama who had never reached a clear awareness either of himself or of others except as routine or impersonal role-players would be led by Shakespeare's visualization of the Lear-Cordelia relationship and by the power of his language to participate in this development and in the fate and meaning of the aged and doomed king. The poet and dramatist can sympathetically enlarge human experience even to include those events and things, e.g., persons, for the description and understanding of which we have initially neither an adequate vocabulary nor the appropriate concepts. Poetry and drama thus offer a means for referring to ourselves or to others and for effecting change by activating or reactivating in us an analogous experience. They often awaken in us by the mimetic means, a new and lively self-feeling and awareness of our own possibilities and destroy old roles and routine responses. Role-players are selves even before they are role-players, and self-knowledge is always latent in us, always susceptible of a more explicit and evocative expression, once the dialectic between feeling and reflective interpretation is effectively initiated. In sum, though we are only dimly and partially aware of the possibilities of self in the flow of feeling at the level of spontaneous interpretation, a selection of these feelings may be evoked by rhetorical techniques and offered for reflective interpretation. [16] Thus the non-objective dialectic is begun. And thus we make our human possibilities explicit in the pursuit of wholeness (cf. §20).

## §10. *Role-Playing as Self-Interpretation*

The self's power of self-interpretation is exhibited on a reflective plane in choosing and playing roles. Roles have also been noted to possess fixed and object-like properties and thus to be unlike selves. Evidently the relation between self and role requires to be made clearer.

An institution, an industry, for example, or an individual, are, among other things, organizations of roles. The need for the division and specialization of labor enforces this specification and systematization of functions. A role is a group of typical acts which are unified by an intention and perform a function. Usually the function is socially useful or at least is defined within and limited by a social structure. Some roles are complex and perform several functions. Just which functions and how many fit appropriately together in order to define a complex role which a single skilled individual can reasonably discharge varies from institution to institution and from time to time. The *Republic* of Plato is one of the first successful attempts to determine the indispensable functions and the relations among them characteristic of a just society and of a just citizen. Being a citizen, however, is only one of the complex roles which a person[17] is normally called upon to play. Such a role is an element in a structure of individuals—workers or other functionaries—which remains constant even though the individuals exercising the roles may change. In the Platonic Republic, for instance, the roles of legislator, executive, and producer are exercised in different ways by each citizen, and these roles persist while the citizens fulfilling them change throughout the generations. Plato was concerned to understand that element which remained constant. But the very fact of constancy of role under change of functionary indicates that the role does not define the functionary himself. Rather the adopted role interprets the self. It expresses within the context and structure of a given society what a self believes itself to be. A role may, then, be regarded as a decision concerning one's identity which is being tested by living it. Thus a professional physicist is testing out the decision: "I am a physicist."

The self who plays roles in a fixed institutional context may lay claim to one identity while playing many and changing roles, father of a family, political entity, factory worker, friend, professional man, tradesman, and the like. The self, moreover, may change his role or function more than once during his lifetime, but the self need not on that account suppose his identity to be changed. Rather he is expressing the many possibilities which possess his one identity. Self and functionary are distinct. We ask, therefore, who or what that identical element is which plays many roles or performs several functions, and changes roles. Since we hold that this persistent element is basically a self, then we would like to be able to point more unequivocably to selves as distinct from their roles.

A few examples of self and role-player will serve to bring home the difficulty in this distinction. In our culture if one's physician dies, then one gets another. The physicianly function, however, does not die. The physicianly function is a role which may be filled by many role-players. A role is something like the invariant in a mathematical statement; role-players are its variants or its range of values. Otherwise expressed, a role is constituted by a definite specialist function. Ideally, individuals in their capacity as role-players are interchangeable with respect to role. In fact, though, some role-players exercise their functions with more skill, imagination, and success than others. Thus, the invariant character of the role is obscured by the personality and qualities of the role-player. We may think of a role as a part in a drama: some actors are more effective than others. But the role survives even the best actor—or the worst. Also some players succeed better than others in expressing themselves in their parts, for roles manifest the self as well as fulfill a social function. But though they manifest the self of the player, they need not be confused with that self, even though upon occasion one may encounter persons who seem to be nonentities apart from some favored role.

The negative proviso that selves are to be contrasted with roles and role relations is always helpful to recall. Thus, unlike role-players, selves are unique. Now, a mark of uniqueness is non-replaceability, and surely we hold selves to be irreplaceable. We like to believe that the woman whose husband dies does not normally replace him with another. Rather, she engages in another marriage with another person. Perhaps someone disposed to object to this distinction will remind me that the term "husband" is derived from the Anglo-Saxon words "hus" and "bonda," literally translated as "house-servant," the reference being to former times when the lord of the manor was in the habit of absenting himself frequently on military expeditions. At such times his lady, left behind and in charge of the manor, might take unto herself a hus-bonda in order to keep the situation under control. The hus-bonda, thus, was a stand-in for the absent lord. Now I agree that this ancient situation may not be unexampled in modern times. Still, I believe it not to be universal. There are perfectly genuine instances in which a wife does hold her husband to be irreplaceable, definitely not one of a number of values substitutable for a social variable. Such a wife would understand what is meant when it is said that a self differs from the role-player not merely in being unique but also in being irreplaceable. Uniqueness is a logical property of a personal self; irreplaceability refers to its value. It will be sufficient to accept irreplaceability as the criterion for status as a self. If a human individual x can be replaced in a given situation by another individual y, then x is functioning not as a self but as a role-player. Replaceability operates as a kind of test of the role function. A self, thus, is an irreplaceable human individual. A role is an element of a social structure within which an individual may function and so reflectively interpret the self who chose the role. Functioning thus as a role-player, the self serves the society within which the adopted role is defined, and he serves himself by coming to know himself explicitly as playing just that role in just that concrete way. In the first way of serving, the individual is replaceable; but not so in the second way.

The difficulty in advancing beyond illustrations,

generalizations, and tests such as those offered here lies not merely in the theoretical difficulty in defining a unique individual, it lies also in the practical difficulty of recognizing when one actually has experience of an irreplaceable individual, of a self. I cite for illustration the records of divorce courts. They indicate that many a married couple have supposed themselves to have entered in on a personal relation only to discover that each was replaceable by another. The personal situation was illusory; their marriage proved to be mere role-playing. This last illustration is informative; it underlines the difficulty of discovering genuine instances, even of instances merely similar to, the personal relation. What can talk about selves mean to someone who has never had experience of self? This question is tantamount to asking how a missing item of experience, here experience of existing persons, is to be produced as well as how further communication concerning this item of experience may be effected. At best we may observe here that the self is that which can play many roles without sacrifice of its own particular identity.

A person need not be aware that he is playing a role. In fact he is often unaware of it; then his performance is uncritical and remains close to the level of involuntary or spontaneous interpretation. Erich Fromm has developed the concept of the social character, or as we might also term it, the social role. This is the character which a citizen is led by a society to develop in order that he should desire to play the particular roles or perform the functions which the society requires for its continuation. [18] One possessing such a character tends to look upon it as completely natural and normal (however idiosyncratic in comparison with others it may be). Other characters, other cultures are immediately deemed to be abnormal, strange, or barbarian by this standard. We are, nevertheless, dealing with a role, though a persuasive one, which the self plays; not with the self. Indeed the self may be so overshadowed by the requirements of a conventional role that the role may go quite unrecognized and unresisted. Of course, all societies exercise their persuasive efforts toward reducing the differences among selves to a manageable average. In addition, some generate so powerful a drive toward the adoption of and identification with certain roles that only the exceptional individual can resist it. Solzhenitzyn is not exactly an average citizen. Later (Chapter VII) the character of such a culture will be considered.

Selfhood, is at the least an individual's power of interpreting himself, both spontaneously and reflectively, by adopting and playing certain roles while remaining unique and irreplaceable. The roles chosen exhibit but do not exhaust the power of the self; they make certain of his possibilities explicit while obscuring others.

## §11. *On the Concept of Self*

Once the experience of self and of its distinction from specialized roles has been induced by the dramatic or rhetorical means or by some other, the problem of conceiving of self may be entertained. The problem is important, for as we have observed the dialectical movement toward self-completeness — the non-objective dialectic — frequently fails to move harmoniously with the objective dialectic and with role involvement. A concept or accurate image of the self is surely a desirable instrument of self-interpretation and a seemingly necessary guide to selecting those roles which are most useful or most in harmony with human possibilities, and with the sense of one's own being. Does, then, the dialectic of self-interpretation conclude somewhere with a definition and a clear concept of the self? Or is it, from its origin to its end, a "dialectical notion" and thus unfinished and undefined? And in the latter instance, how should this less than precise image of self be expressed?

As touching an answer to the first of these questions, I must admit to pessimism. It is not difficult to define a role. For instance, the role of physician is well defined and fixed by state board and medical association examinations. Passing such examinations entitles one to practice this art. Similar criteria may be found for very many other roles and functions. The self, however, is different from a role in that the self usually practices a collection, an indefinitely large and varied collection, of roles and functions. It is safe to say that hardly any two persons practice exactly the same collection of roles and functions. On this ground alone, any person might successfully claim uniqueness, but as unique he could not be classified and conceptualized as one of a group of similar individuals. The possibility must, therefore, seriously be entertained whether the self can be conceptualized beyond the innocuous generalization that it is a something which practices many roles. In any event, still another considerable difficulty must be overcome by anyone seeking to develop a concept of self. Can a self fully reflect itself in a concept? If so, then that self-reflection must be an item in the concept. The ancient problem of self-reference seems at the outset to undermine with its infinity the attempt to produce a clear concept of the self.

The sense in which I earlier tried to point out that a person is unique included an additional special trait, that which is involved in the personal relation and which I termed irreplaceableness. Now, I can discover no possibility of forming a concept adequately descriptive of person in this sense. One reason for this inability is quickly expressed. The very condition for forming a concept requires abstraction from the differences among individuals, and this requirement entails abstracting from or ignoring the irreplaceableness of a self since in this context irreplaceability refers at the very least to the sum total of differences which render an individual unique. Thus, this condition entails ignoring a primary element in the constitution of persons and so guarantees overlooking their concrete character. I am here merely reiterating a view, widely current since Aristotle formulated it, that the individual is not definable.[19] One defines only classes, and classes are collections of similar, i.e., non-unique, elements. The point is not refuted by citing instances of classes having only one member if the member possessing the class property is also an individual unifying in itself indefinitely many other properties. For if the person is, at the least, a unique structure of an indefinite number of roles, to pick out some unusual or role-related property (e.g., a way of speaking, or a thumb print) and to use it to classify him in a class apart, is scarcely a significant move as respects the present problem. Nor is the point refuted by pointing to a class formed of all unique individuals, for this last is a class only in a Pickwickean sense, having no explanatory or descriptive value.

Is there, then, no means for referring to the self in a general and genuine manner? Perhaps there is. We have accepted the tradition, hardly avoidable, that man is a being possessing language. Also we have observed that a poetic and dramatic use of language performs a necessary function in the non-objective dialectic by which the self develops its self-interpretation. This use of language provokes the feelings and attitudes which are necessary to initiating and continuing the interior dialectic. Perhaps this use of language can be shown to point still more clearly to the self. Let us, then, consider certain traits of poetic language.

It is fairly common at the present to hear that the meaning of a word or of a string of words is to be discovered by examining the way they are used. And the basic unit of use is said to be exemplified in the declarative sentence, e.g., "Cordelia was a blond" or "Shakespeare died in 1616." Now, while these opinions about language may be defensible if they are referred to ordinary speech—that is, to the kind of communication necessary among role-players, —they may not apply to all language used in communications among persons. Indeed, since the sense in which selves are said to exist is just the sense in which they are said to be irreplaceable and unique, they do not, as I suggested, belong to classes, and hence they cannot be managed as if they were members of well defined collections of similar elements, e.g., people having blond hair or people who died in 1616. If Cordelia belonged to one of these classes, then she was not the Cordelia discovered by Lear on the Heath. Nor is it helpful to observe that selves are designated by proper names, for proper names are ambiguous. "Cordelia" pronounced by a professor of literature and by King Lear may refer respectively to a role, i.e., to one of the dramatis personae of the play, or to the person herself. Nothing new is said thereby concerning the way in which these two usages achieve their different kinds of meaning.

It is of interest, however, to observe, as Plato in the *Cratylus* observed, that words in ordinary usage are instruments, functionaries. They are linguistic role-players whose roles are defined by commonly observed syntactic conventions. They are citizens-workers within a relatively fixed linguistic city-state. Only so are they serviceable either for daily usage or for other specialized and more exact communications. In poetry, on the other hand, syntactic and other conventions are not fixed; the poet is commonly allowed an exceptional freedom to alter syntactic patterns and semantic usage. Consequently, his words tend to shift in unexpected reference. Although not many poets have used syntactic and semantic license so openly as (say) James Joyce, all have laid claim to this freedom. This poetic liberty with and freedom for language suggests that not ordinary communication but something else is determining the poet's choice of word and word-structure. If poetically used words are instruments in some sense, they are not exclusively so; at least they are not merely citizen role-players in a linguistic city-state. Their additional character has often been described as disclosing the unfamiliar, precisely that which ordinary language is not designed to do.

The relation of this character of poetic language to ordinary discourse is indicated by taking note of the function of the declarative statement. Note that declarative statements — "Cordelia was a blond," or "Shakespeare died in 1616" — are answers to definite questions. What color was Cordelia's hair? When did Shakespeare die? They bring such questioning to an end by providing definite answers. If declarative statements do not answer such questions, implicit or not, they are useless. Thus, there is a sense in which the interrogative is prior to the declarative; the declarative — the stated answer — presupposes the question. Now this prior interrogative discourse is suggestive of the poetic. As a question seeks to elicit more than has been said, so poetic language strains to say more than ordinary

language can compass. Notice of this strain is expressed by the
remark that the poet seeks to express the unfamiliar; he seeks to
disclose something perhaps always there but hitherto
unrecognized. Thus, in Shakespeare's hands, Lear for the first
time recognized and accepted himself and his fate, though his
fate was his own prior to royal station and to fatherhood.

How often has it been written that the poet speaks as if what
he says were said for the first time! For just such a reason poetry
must destroy ordinary language; hence, it uses extraordinary
syntax and takes liberties with semantics in order to draw new
possibilities of meaning out of the words of daily discourse.
Positivists have correctly observed that poets speak in pseudo
sentences; poetry tends to strain or to destroy the moulds both
of daily speech and of precise scientific discourse. For the poet's
experience, being experience of the unfamiliar, he is impelled
to refer to that which is hitherto hidden and has no name.
Hence the poet is reputed to be the discoverer and the name-
maker. He is the originator of language. But he does not usually
manufacture names out of whole cloth; rather, by the use of
metaphor—as Aristotle expressed it—he gives one thing the
name which belongs to another.[20] To correct, as it were, for his
"error" he seeks to give the name in such a way as not to terminate but rather to preserve the sense of interrogation or
wonder. Metaphor used poetically always carried with it a
question concerning its meaning. What is that strange light
which never was seen on land or sea? Who is he that "doth but
slenderly know himself?" even though he has been living with
himself for 70 years? Where are the eternal edicts of Zeus,
invoked by Antigone, which are present in the world yet are not
of it? Poetic metaphor, like the poem itself, is never more than
an obscurely achieved meaning. It is a feeling of significance,
an implicit questioning, an invitation to the reader or a hint to
the listener to complete the meaning by exercise of his own
prophetic and sympathetic vision.

Revolution, I suppose we may say, is the poetry of politics;
conversely poetry seeks to revolutionize the world. It invites the
mind on a quest, seeking to question the customary and to

deprive us of the familiar. It is both a gift and a theft; indeed the theft is the condition of the gift. Just this threat of stealing away the familiar, the comfortable, undermines the *status quo* within which ordinary language moves and has its being. Little wonder that poetry is excluded from the life of daily business; little wonder that the poet is an exile from the city and an abomination to the analyst.

I summarize these remarks about the poetic *logos* by saying that it is engaged in the disclosure or origination of values. This is a remark which sometimes or other almost everyone has heard, understood, and perhaps believed. In repeating it, I think to touch upon a view, even upon an experience, which is quite common. I want, now, to use it as a basis for indicating the nature of the self. I want to evoke this nature by use of a poetic analogy: this analogy holds the language of poetry is to the functional language of daily usage as existing selves are to role-players. Poetry, the analogy says, tends to break out of the context of ordinary language and cannot be defined within it; similarly selves and personal relations tend to escape the structure of roles and social functions. The self's possibilities are always more than their initial interpretation in feeling, and this first interpretation is always more than the reflective interpretation in role.

The relation I have in mind is rather like that between the daily tasks of a society "getting and spending" in meaningless routines and the ritualistic and festival celebrations which tend to bring the participating members into revivifying touch with the values and symbols inherent in their culture and world. Like such a ritual community, like a poetic work, the self is the possibility of indefinitely many interpretations. It is the source of roles which seek to interpret their origin and can often do so in new and original ways. Hence the self is not defined like a chemical element is defined, as a definite structure, exhibiting an endlessly repeated chemical function. Rather, it is "defined" by alluding to the power of eliciting and developing new and perhaps unpredicted possibilities. The self interpreting itself is related to its possibilities as the poet is related to his language.

Both self and poet function as makers; both seek to elicit from all their world and their history have delivered to them a new or renewed work. The self is the being having this capacity for self-definition and can sense this power as the obligation to pursue insight into its own possibilities and to release their powers. In exceptional instances the self may even engage in radical alteration of its whole response to its total world, as a young Athenian aristocrat became a philosopher, as Saul became Paul, as a petty bourgeois banker became Gauguin. It must also be added that contemplation of radical self-change reveals a threat: the radical change may be from an Austrian house painter to a Hitler, from the innocuous to the immeasurably guilty. The possibility that each of us should hold within himself such capacities is enough to direct attention upon the meaning of guilt and upon the limits within which self-change needs to be confined.

For the moment we note only that poetry, operating to disclose values, is valued for itself. In consequence, it lies always outside the circle of practically or theoretically justified statements. It reflects astonishment upon the revelation of something strange, unexpected, and incompletely grasped, strange at the least just because it is not manageable and analyzable in everyday language. So existing persons are, as Kant remarked, valued for themselves alone; they are outside the utilitarian sphere where men must play roles and use each other as instruments. There is always more in them than existing institutions and concepts have used or understood. This notion of the unique, irreplaceable, and creative self, suggested by the poetic function of language, seems eminently well fitted to express the Western world intuition of the indefinitely remote completeness of man, a completeness from which the existing self is indefeasibly alienated, but a completeness which the self can apprehend by means of its strange power of standing away from itself in self-awareness and self-interpretation. By way of the same power the self recognizes its impotence to attain its completeness without a strike of genius or some other extra-individual power.

No doubt, then, when we address an existing self, we do so poetically, though we may do so only by a gesture or by mention of a single name. I believe thus to touch upon an archaic experience which, however, is not uncommon and is available to all. "Self" refers primarily not to a value for a social variable but to a unique, irreplaceable, and creative individual who can be apprehended as such by recreating, through the dramatic and poetic means, the experience of such a one. The possibilities of self are brought forth and made manifest by the insight of poetic vision guiding action, as the sun brings forth and illuminates the hidden richness of earth. But as the sun has its own position and movements and is limited thereby, so the poetic *logos* is limited by the world one inhabits. Self is an achievement, reached by pursuit of the concrete truth of oneself. Symbols, concepts, and handy phrases may be, however, useful pointers toward this achievement since, if they are adequate, they provide some sense of the complete self, of its limits, and of the world in which its completeness might be pursued (cf. §27). To the extent that the philosopher is concerned to determine the nature of the self, we must, therefore, observe that not the laboratory but the theater should be the source of his data.

It must be added, though, and the theater of any language will verify the observation, that roles need not be played in a manner constantly to elicit new possibilities of the self. Often they are played in a stereotyped manner. Then the self identifies itself with a role and bends every effort to avoid bringing this identification into question. It sidesteps the periodic deaths which should mark the human career (cf. §§22, 26). One of the possibilities of self is blindness to its own possibilities. Radical change, nevertheless, always remains a possibility; hence, the self cannot be conceptualized in a clear and objective manner. A concept of the whole self is more like a question than a definition. At best it is a dialectical concept (§9).

Furthermore, as poetry is destructive of ordinary language, may the self not also be destructive of the role-player? In fact, poets are savages in the brave world of our decorous and precise language games. Similarly, I suspect, persons or selves are strangers in a well-ordered society devoted to industry, business, and technology. They are poets in the Platonic city. But not for this reason alone need they be banished or put to death. Let this section count as a modest effort to recall them in person. [21]

# CHAPTER III

# THE PRIMARY OBLIGATION

## §12. *Obligation and Human Nature*

Given a man who refers to himself as a self without quite knowing what a self is, what will be his character? A more modest question: what should we expect his primary motive for action to be? Not unlikely such a person will feel a lively sense of obligation to discover his powers and to make his being manifest. Perhaps he reasons in the following way. Since I am a man, I desire to avoid guilt. But only if obligations are met can guilt be avoided. Here an obstacle prevents drawing the conclusion, for obligations are a function of what man is, and what man is is unknown. The last chapter concluded that the self is a structure of possibilities which are initially limited yet are rich, as a poetic writing is rich in possible meanings. But who can say in which direction the most human meanings will be found? Consider how differently human being has been envisaged from one time to another or from place to place. The first obligation, therefore, must be to discover the profitable direction of human development. Then one may set about making this being manifest in action.

The obligation to know and to become one's complete self seems clearly to be primary. But others are scarcely less important. Generally, it would be admitted, one has a similarly basic and common obligation to other persons, for instance, to minimize one's own interference in their efforts to know and to complete their own being. No doubt one also owes a primary debt to nature, as those interested in the various ecological movements are constantly pointing out. These three kinds of obligation arise from the threefold relation, the relation to self, to others, and to nature, which is constitutive of world (§4D). Thus, these primary obligations are those which compose the obligation to the world and are common to all who inhabit it. I shall be concerned here only with the one of these three which is most germane to this essay; this one is the obligation to oneself. For this chapter it will be *the* obligation.

Let us ask, then, what is the meaning of this obligation? Who is so obligated? Why is he thus bound? To whom and for what? I propose to follow these questions through several levels: the feeling of obligation (and guilt), culturally imposed obligations, reflectively defined obligations, and obligations of the questioning self.

A well known recognition of the primary obligation is given its setting within a prison. Crito awakens Socrates at dawn, and they quickly fall again to debating whether Socrates was obligated to his friends—and against the legal decision of the City—to flee Athens, a possible and—as Crito saw it—attractive alternative to drinking the hemlock (*Crito,* 48B; 50A). Socrates argues that he feels obligated to the Athenians as if by contract to carry out their wishes, a contract implicit in his relation to the city which had fathered him (48B). He concludes that, in order to remain himself and to continue the pursuit to which his life has been dedicated, he must fulfill this contract. This contract seems toward the end of the dialogue to become a relation between him and the Ideas, those intelligible objects in whose image both he and the city desire and ought to desire to shape their characters, but into whose nature he has but fleeting and imperfect glimpses. Socrates is, thus, motivated by awareness of obligations to himself, to others (the City), and to the intelligible natures. His obligations arise, thus, from what I have called the world. His motives, though, are Athenian and differ from the motives characteristic of a man of Western world culture. These differences, however, will not be the point of emphasis in this chapter. We shall be interested in reaching a general view of obligations to the self. For this purpose the Platonic example is typical.

This Platonic illustration, moreover, suggests a convenient formal mode of expressing this kind of obligation. As Socrates is obligated by the implicit contract between himself and the Athenians to obey the injunction of the judges in consequence of his filial relation to the city which had fathered him (and finally to the Ideas or Laws themselves), so more formally we may say that A is obligated by B to perform C in consequence of some reason, D. Thus a physician might argue: since I am a physician, I am obligated to my patient to effect his cure in consequence of my professional position and legal regulations, a position and regulations which reflect the world which enabled me to become what I am. The genuine professions have always contended that their standards are voluntarily accepted and self-imposed; hence, legal regulations are invoked only in exceptional instances. Normally holding this professional status, precisely that in our culture which makes a man a physician, is that which imposes the obligation to effect the cure of the patient. To fail to cure the patient, when a cure was possible, imparts guilt to the physician. Whether this guilt be thought of as a set of feelings, a state of conscience, or a collection of reasons does not matter at the moment; such guilt is, in any case, a blot upon the scutcheon of the physicianly conscience. If there is any force behind the imposition of physicianly standards, probably it is the force of this guilt. Thus, our physician really argues: if I am a non-guilty physician, then I am obligated. But surely in normal instances a physician is a non-guilty physician. Socrates seems also to maintain this position. The reasons giving force to his obligation, he says in effect, were built into his character by his education in the Greek world and in Athenian culture. They make him what he is. Flouting these reasons would impose a guilt far greater than the legal guilt imputed him by the Athenian judges. My point, then, is this: when the A in our formula is an autonomous self or something like one — e.g., a self-disciplining profession — then the D in the formula is held to be included within the A; it constitutes A's character-structure. The formula then reads: If x is an A, then x is obligated to B to perform C (even when the performance of some non-C is possible and in some ways more attractive). The obligation which is of interest in this chapter therefore, is the relation holding among A, B, and C.[1] Investigation of this relation should throw light upon the A, the self, which can enter into such a complex.

The problem of obligation is the problem of understanding and of giving appropriate meaning to the elements and relations of the formula drawn from the Socratic example. This formula was expressed hypothetically, a convenient form of reference for those who perform some special role. Not every one is a physician, but if one is a physician, then one is obligated thus and so. If one performs another role, then the obligations are accordingly different. Are there, however, any obligations which are laid upon a person just because he is human, whatever his special role? If so, then they would be laid upon him unhypothetically. There would be no question whether x is an A, for all these x's are persons, and persons are necessarily human. But all are involved in the question whether there are any specifically human obligations, and if so, what these are. It is clear, I think, that in Plato's view the debt which persuaded Socrates to lead the life he did and to accept his death from the citizens of Athens is a common human obligation. The especial virtue of Socrates lay precisely in his recognition of this obligation and in the questioning way of life which made this recognition intelligible to some others. When we speak of this Socratic obligation we are making Socrates the A in a quite common obligational complex. This Socratic obligation is the kind — I shall call it *the* primary obligation — with which this chapter is to be concerned, for it involves the nature of the self in a quite special way. The power of this obligation, it should be noted, is such as to lead one upon occasion to sacrifice one's self. It suggests a close relation to a profoundly held conviction which determines the meaning of human life itself.

§13. *The Older Self and Feelings of Guilt*

The self has been identified with its possibilities, possibilities which are initially exhibited as feelings. A very important class of feelings are those associated with obligation and guilt. Consider first the sense of obligation. These feelings are spontaneously interpreted as valuations of a certain kind. They are, as it were, anticipations of value[2] which may be experienced in some future event. The feeling of obligation, then, is the anticipation of a value.

Such valuations can become motives to acts. As it might be formally expressed, A performs C because of certain valuations. Now to be obligated is at the least to choose and adopt a certain class of valuations; more than that, it is to feel oneself committed to the pursuit of values of a certain kind. This commitment (to B in our formula) is not a matter of the comparative indifference nor of the mere preference with which many value choices are made. Rather on its positive side, this class of valuations is interpreted, usually quite spontaneously, at least on the part of persons well attuned to their culture, with unique force. In addition, this valuation possesses a rather peculiar negative power. Violations of obligations are felt to place the violator in dire, if frequently vague, jeopardy. This peculiar force of obligatory valuations — both on its positive and negative sides — is difficult to understand. What makes us feel obligated?

Let us say that A holds himself obligated to B to perform C if the consequences of the non-performance of C is a sense of guilt in A.[3] Socrates offers an example; if we can trust his statements, his decision to flee prison in order to avoid the sentence imposed by the court would have been followed by (among other things) a lively feeling of guilt. He would probably have felt a loss of self. An empirical test is suggested by which obligations (valuations substitutable for C) can be detected. And in many areas and eras, especially religious eras, this test has been accepted. "Let your conscience be your guide," it is confidently said. Here conscience is experienced as an easily recognizable feeling. Likewise, some recollection has remained of the etymological sense of "conscience." Conscience in this sense refers to knowledge (*scientia*) which is had in common with someone else. In the traditionally religious instance this someone else is said to be God, specifically the father-God, who reveals the standards of human behavior and who punishes and rewards accordingly. This God is the B in the formula above, an authority who ties obligations onto men by the thongs of a feeling of obligation and guilt which he mysteriously but systematically elicits in his creatures.

A little travel in foreign lands where men having a differently trained conscience live under other gods who prescribe different standards has often been sufficient seriously to undermine the conscience of a person. Likewise it is disturbing to the confidence sometimes placed in this bond between god and man to observe the sense of intolerable guilt which leads the psychopathic personality to confess to a murder which he never committed or which leads the religious devotee to believe he had crucified the Christ within him because he robbed a neighbor's pear tree. If the pangs of conscience can and do vary in this fashion both with the social and individual milieu, then they may indeed be a misleading clue to the nature of obligation and to the authority which imposes it. But this familiar conclusion is too quickly drawn. There may be an element in common among many of the occasions of feeling guilt which, if it can be disengaged and expressed, will prove to be helpful.

A step toward disengaging this common element can be made by comparing the sense of guilt with certain other feelings to which it is closely akin. It is similar to and is sometimes confused with embarrassment, for example, or with a sense of shame, wounded pride, or the confusion which follows upon the commission of a mistake. It will be profitable to contrast guilt with the feeling which accompanies the recognition that one has made a mistake. We approach this matter by recalling a psychological theory concerning the development of conscience which is held to account for the sense of authority with which obligations are imposed; we refer to it in order to account for the diversification of the sense of guilt from other feelings of the social being.

Freud, and others as well, have concluded that conscience is acquired in consequence of the operation of the natural tendency, which he posits, to imitate or to "identify" with certain others. The child imitates (introjects) other people and thus absorbs them or aspects of their behavior, into his own personality (cf. §20). Among those whom he imitates in this manner are other persons who are engaged in observing and criticizing him. He then, taking himself as object, mimics this criticism of himself. Thus, eventually he comes to acquire an observing and critical attitude toward himself and, like his father or other authority, to punish himself for that which he has adversely criticized. Primarily, he introjects or incorporates his father or father-role and is said in consequence, to acquire a set of internal standards or a conscience which actively opposes and controls his primitive impulses. The admission that some act or omission is to be adversely criticized is registered emotively as a sense of guilt. To the child, who makes no fine distinctions and tends to react wholly or not at all, the spontaneous interpretation of guilt feelings is not related so much to a specific act or error as to the self as a whole. The child does not say to himself, "This action was wrong and should be corrected." Rather he says, "I am evil and deserve punishment." The child makes no distinction between awareness of a mistake and feeling of guilt.[4]

Obviously his feeling is misleading and cries out for further interpretation. Freud turned his attention upon it. He found the infantile interpretation of guilt feelings to be nearly universal and concluded it to be the source of a paradox of our culture. For if one attempts to satisfy his often murderous impulses directly and forceably, then he is punished by society, and if he voluntarily frustrates them in the interest of gaining the love and approval of others, then he still harbors the wish to gratify them at the expense of others, and hence he feels guilty and punishes himself by further self-inflicted frustrations in order to rid himself of these feelings. In either event he is punished. Freud thinks that civilized life will become more and more complex, and the occasions when he must choose between criminality or self-frustration and self-punishment will become more and more numerous until at some point the tension between impulse and conscience will become too great for him to bear and he will be torn asunder.

This dire prediction may not be the only alternative, for note that the incorporation of the father as the interior power and ruler has already set up in the child an internal model of the family of which he is a part. In the movement to adulthood, diversifications of this interpretation occur in consequence of the introjection of the roles of other and differently functioning individuals. One might build a case for supposing that not only the ruling father but also the hunting and fighting older brothers as well as the agricultural and weaving women acquire their representatives within the psyche of each normally developing child. A society passes through the process of ruler-selection, division of labor and specialization for protection and production, so the individual who initially tends to incorporate the social authority will also find himself incorporating other functioning roles of the social structure. Thus, he will come to mirror in his personality the functional divisions of society. He acquires by this mimetic communication not only an internal authority who imposes an order and punishes its infringement but also becomes aware of a rich psyche, able to accept multiple functions. As society grows in diversity and clarity of its organization, so we must suppose, does the individual. At the least, he develops not only an internal ruler, but also internal artisans who use skill to produce satisfactions for needs — more or less in harmony with the internal rule — and likewise internal protective and executive roles which mediate between ruler and artisan and protect the whole from internal as well as external injustices.

The upshot of this development is that the person need no longer feel his whole self to be guilty for any chance error, as these are measured by the complex and graded civilized standards, but on some occasions he need feel only that something went wrong with some role which he was playing or with a subordinate aspect of his functioning. In effect, his feeling of a pervasive guilt consequent upon admission of error is dispersed over many functional parts or roles of the self, just as the tasks of providing for society's needs are distributed among many specialists. Error thus becomes less personal and easier to bear; compensation for it is more varied. One important product of civilization, perhaps a saving element, may be just this diversification of error and guilt, in relation to the more complex self, and differentiation in the emotional reaction to error. Not all errors, only some, are violations of obligation. Which errors are these?

Some errors have always been connected not merely with a role or specialized function of the self but rather with the very identity of the self. Let us express this important distinction as the difference between sense of mistake and sense of guilt. One feels guilt when the acknowledged error violates or conflicts with one's identity. Or again, one feels guilt when the acknowledged error is associated with the self's violation of his internalized ruler, the father within him, rather than when the error is attributed merely to some subordinate aspect of the self. But if the actor's identity is not called into question by the error, then he is aware merely of a mistake.

It is sometimes difficult, as we have seen, to make a clear distinction between the self and roles which one plays; again, I make no claim to define the self. It is sufficient to note that the self's relation to the parents and parental authority is certainly a basic relation, one without which the person would not be. Naturally this relation figures fundamentally in the psychic economy. So intimately does it function, I suggest, that one's sense of self is often identified with the internalized father-authority. It may be, as we noted, that any occasion of feeling guilt is connected with the atavistic impulse to absorb him, his power and his function, within one, and through this violation to become the father-god; any impulse or act which reawakens the feelings connected with this original impulse is felt as guilt. Thus the sense of guilt is felt when the very self is taken to be the culprit. This self, the A element in our formula, is the element

common to guilt feelings which lends to them their peculiar
force. The sense of guilt is the feeling associated with a loss of
self. This is the feeling which lends power to obligations. When
error is attributed not to the self but to some aspect of the self,
or to some role—to the self as soldier or as artisan for example—
then one is aware not of guilt but of a sense of mistake, the
uncomfortable feeling of an error in technique. Thus upon
being interpreted, the initial global sense of guilt is dif-
ferentiated, and the feeling of guilt, accordingly, becomes more
sharply specified and directed. This new sense of guilt now
demands further interpretation.

## §14. *The Hypothetical Self of a Culture*

Up to this point I have been concerned chiefly with the
feelings experienced upon occasions when an obligation is or is
thought to be violated. Now it will be desirable to free the
concept of guilt from its origin in this feeling and even from the
mytho-psychological scaffolding which was needed initially to
specify and to relate the sense of guilt to the self. We need
reflectively and critically to interpret these more sharply dif-
ferentiated feelings of guilt and obligation. For clearly, to know
that people are apt to feel guilty upon certain occasions does not
provide what we seek, namely, an answer to the question why we
are obligated to avoid guilt by performing certain acts or
adopting certain valuations. All a psychological account can do
is to tell us that by the operation of an effective mechanism a set
of possible choices is in fact placed before us in a powerfully
persuasive manner. But an account of this sort does point to
relevant data. We may now move a step further toward an
account of this obligation by the aid of the notion of the ideal
self indigenous to a culture.

The notion of ideal person is implicit in conscience. One feels
oneself responsible to the father or the authority component in
conscience to become precisely the person which is not yet: the
ideal person. This ideal person may not be clearly described,
but, buried under tradition, convention, and formularies, it is

nevertheless an essential part of every person's cultural inheritance. Occasionally this ideal person is hatched from its cocoon of convention and expressed with some care and even criticized. When the ideal person implicit in a family, tribal, or cultural tradition is rendered explicit, I shall call it the cultural or hypothetical self. I call this self "hypothetical" since it can lay no claim to being absolute but is defined within a tradition or a culture. The notion allows us to define actual guilt as distinct from the sense of guilt. When A holds himself obligated to B to perform C, and the failure to perform C, when C was possible, is followed by the recognition of a specific disparity between the current self (A) and the cultural or hypothetical self (B)—a disparity which would not have existed had C been performed,—then A has recognized actual guilt. The sense of guilt, then, is a feeling which is intellectually interpreted as the awareness or recognition of this disparity. This definition takes the emphasis off feeling and emotion, a move which is desirable since guilt feelings may be absent when actual guilt exists and vice versa. The *Crito* provides an example of actual guilt. Socrates, endeavoring to lead the examined life, trespassed against the reigning view of man and his obligations and was, accordingly, found guilty. The fact that we can see, being led by Socrates, that the Judges were motivated by self-interest and a cowardly anxiety provoked by Socrates' life of questioning does not alter the actualities of Socrates' situation. Our adverse judgment of the Judges is a rejection of the society which they represented quite well. After all, Socrates accepted the judgment, but he accepted it in order to make manifest the state of Athenian justice as measured by a standard which was then foreign to that society.

The B in our formula has now been identified as the hypothetical self, the clarified and criticized cultural ideal, the perfect father which the son as a man of his culture, ought to aspire to be. The standard by which those valuations which are to be called obligations are selected and measured is the hypothetical self. Thus, obligations are self-imposed. Whatever the psychological mechanism by which we are persuaded to accept obligations, it seems to be obvious that no external authority or force can impose them. Obligations, if they figure in morals, are a class of choices, and only the free and responsible person can make his own choice.

The transition from the preceding section which we have made is the following: as the sense of guilt is to the feelings most intimately associated with the self so actual guilt is to the cultural self. This transition provides us with a definitional measure which, relative to the culture whose standards are being used, distinguishes between genuine or actual guilt and pathological guilt. It enables one realistically to criticize oneself by comparison with an explicit and widely shared ideal. The point emerges again that guilt, whether guilt feelings or actual guilt, is intimately associated with the self and is aroused when the identity of the self is brought into question, rather than when some role or minor engagement is involved which is not thought to touch upon the identity and unity of the self. (This remark does not hold good, however, in those instances where a self identifies itself wholeheartedly with a privileged one of its roles. To this point we shall return later, §37.)

The definition which has just been offered is quite general, for it can be shown to be meaningful in any cultural context. This generality obtains when the notion of hypothetical self (the B in our formula) is taken as a variable which may be given content by indefinitely many doctrines, springing from different cultures in different worlds. The question inevitably arises concerning the limits on the range of the variable B. I suspect that this question is inevitable, for it bears on what seems to be a persistent longing to transcend cultural and doctrinal limitations. If obligations are, in the last analysis, self-imposed, what, men have always desired to know, is this self which imposes and accepts obligations? A considerable part of the preceding chapter converged upon the conviction that any concept of the self must inevitably be distorted in consequence of a native inability to see or to understand one's self in a final or complete sense or in a universal context. And one recalls Socrates' injunction to know oneself together with his protestation of ultimate ignorance. Then there is the Christian recommendation to imitate Christ, who at the same time is said to be a mystery beyond man's understanding. Finally, there should be recalled the conviction that man is primarily the drive to transcend any concept of his essence, any identity which he may adopt. Obligations to the ideal self, then, are incorrigibly infected with a kind of uncertainty in virtue of which any act, however virtuous seeming by cultural standards, may turn out to be inappropriate, a thing of guilt, an evidence of a basic fault. We remember again the judges who presided over Socrates' condemnation.

This important point may be made in other ways. I have already indicated the ambivalent attitude in which obligations are held by remarking on the identity of conscience with the father; but the father is a powerful and a dangerous being to challenge and internalize. This being, however, is interpreted as the ideal part of the self, the ineffable self which one both fears and strives to become. The thesis of the incomplete intelligibility of the individual gives expression to the suspicion that any given theory of the individual will distort him, perhaps tragically, in the direction of the aspect within which the theory is developed; and yet, paradoxically, intelligent action can scarcely go forward without some vision of the self that was and is to be. Not the least important presentation of this radical but unavoidable self-uncertainty is offered by the Greek doctrine of tragic hybris. Indeed, reflection upon this weakness suggests a property of human being which goes beyond cultures and characterizes the world itself.

## §15. *The Complete Self and Ontological Guilt*

Although the hero of a Greek tragedy was a superior person, it will be remembered that he always made a decision about himself or the office his duty required him to perform which was out of character or unrealistic in that it demanded of him greater knowledge and skill than he actually commanded. He felt himself to be virtually omnipotent in some respect, but upon being tried by fate he was forced to recognize that he had misidentified himself. Hybris is the tendency of the hero to mis-interpret himself and to attempt the deed which he cannot perform or to assume the identity which is not his own. It is a *de facto* blindness. Christianity and later Western cultures carry this view of man a significant step further. They detect an *essential* blindness, a basic deficiency, in the human being by reason of which he cannot but fail in reaching his highest satisfactions, unless special power or direction be accorded him from beyond himself. One's highest obligation may be to become an ideal self, but efforts directed toward self-realization are so consistently frustrated that one is led to suspect the presence of some ontological impediment in the path to self-identity. As the myths of man's fall insist, it is as if one were already guilty, being human. In being born, one attempts to take possession of that which one cannot be said to deserve and which one cannot own: namely, one's existence and fate.

The psychological myth alluded to in section 13 suggests the same point when it speaks of the rebellious sons who try to absorb and to become their father. In fact, though, they are not their own father. But the over-weening tendency to regard oneself as the center of the universe and the source of one's being persists. Consider, for an instance, the philosopher who flatters himself that he has incorporated and expressed reality itself in his systematic formulations. Or consider the moving spirit of our own times, so confident of the progress achieved by one generation absorbing and building upon another, moving to the super-man technician at its summit; it suggests nothing so clearly as the foolishness of the tower of Babel whose basement is the earth itself where reserve coal, oil, and uranium ore are conveniently stored for use by the developed countries.

Another and less psychological way of describing this defect follows from the kind of limited being which is man's (cf. §4D). As a man's movement in space is confined in any instant to some definite direction, so his attention is limited to some definite aspect or finite group of aspects of the object to which he gives his attention. In general, his powers are not unlimited; to be

effective they require concentration, focusing, upon some
particular. If the sun, sending its rays out in all directions is the
symbol of transcendent power, the symbol of man's limited
power is more like a search light which illuminates only some
portion of its objective, never the whole, even when that objective is oneself. The whole object, or the object viewed from
another perspective, may be quite different from it when viewed
from the given perspective.

Again, we recall that being, the power of self-manifestation,
is always a directed power. An ontological condition of man's
being is directedness in a specific direction, not in another. Man
is always in a finite world. Those modes of being, tasks, and
values away from which a man's power is directed may be
perfectly valid and desirable, as much, perhaps, or more so
than those which were chosen. To direct oneself away from
them is a kind of denial or rejection of them. To choose one
value is necessarily to discard or to ignore an indefinite number
of other possible values. If, then, to reject a value marks a
defect, then to reject value in consequence of the being one is
marks an ontological defect. Hybris results from a failure to
grasp the limited kind of being which one is and leads one to
interpret oneself as the being which one is not. The Christian
doctrine of the fall of man affirms that a man cannot but
misinterpret himself. He cannot but mistake himself for an
impotent or an omnipotent being. In the Western view, man is
so indelibly marked by this ontological defect that even his most
heroic efforts cannot compensate for it.

The greatest hero remains alienated from the complete self.
The whole and transcultural self, this tower beyond tragedy, if
only one knew where and what it were, might be used to take
the measure of any hypothetical self or cultural ideal—in
particular of the technological culture. And then our
obligations, the C in our formula, could be determined. But
this self seems only vaguely to be dreamed in the inspirations of
the prophet and related in stories about the Divine Man,
Nirvana, Sainthood and friendship with God. With the
Christians it becomes indefinitely distant. The Romantic

philosophers despaired of it while pursuing it. The Positivist puts it down as a ghost story.

If our obligations could be justified only in terms of inspiration and poetry about the whole self and its ontological defect, they could easily be challenged. Perhaps only language used poetically or metaphorically can point to the self; nevertheless, it remains true that poetry about the complete self does not offer a convincing answer to the problem why one is obligated or why one is guilty. Also anyone might admit that all men have always held themselves obligated to do their duty and admit, further, that he is a man; still he could quite logically ask: "Why should I choose to obligate myself thus?" To point out that by doing his duty he avoids or minimizes guilt and becomes his culturally approved self or even his whole self will merely force him to put the question again.

An easy issue out of this impasse is to assume arbitrarily some one obligation, some one imperative which will direct choices. If the assumption is aptly selected, one's ethic may still be saved. And this procedure is susceptible of a logical defense. First principles appear to be of the nature of assumptions; and if there are to be imperative conclusions, at least one imperative principle must be assumed. This procedure, however, is too luxuriantly fertile in ethics, for it can be used to justify any set of so-called obligations whatsoever. Accordingly many philosophers have been dissatisfied with it and have sought to show in some fashion that the primitive obligations which they assume ought to be assumed. Usually they argue that among the many imaginable hypothetical selves one of them ought to be regarded as closer to the whole self because of certain rather persuasive qualifications which it possesses.

Earlier the classic tradition, in interpreting man as the rational animal, sought cleverly to define its hypothetical self in such a manner that it could use this definition not merely as an ethical primitive but as a justified or non-arbitrary ethical primitive, one, moreover, which could be counted upon to lead directly to self-knowledge. Perhaps the rationality (*logos*) of a person is only one of his roles or aspects and perhaps, in identifying himself with it and developing primarily the potentialities of this one aspect (as if the individual were or were to become completely intelligible) is a tragic decision. And yet one must admit the peculiar and forceful persuasiveness of this identification. Without development in just this regard, self-discovery would not be possible, for awareness of the tragic hybris, or recognition of the gap between the existing self and the hypothetical self as revealed in the trial by fate, is precisely a rational recognition. Without this use of rationality no distinction between the currently existing, the hypothetical, and the whole self could be recognized or expressed. To put the matter succinctly, it is only by interpreting himself primarily as rational that a man can find out that he has misidentified himself, if he has done so.

It is easy to see that the traditional convictions about the primary obligation of each man refer back to this identification of himself as rational. The primary obligation of each person is to pursue the being of himself through the rational process of Socratic self-examination: hypothesis, trial, insight and reformulation of the hypothesis. This is to say, a man is obligated to pursue the truth about himself by way of the non-objective dialectic. Inevitably the recognition of a disharmony between what one actually is and what one naturally and more happily might have become cannot be accepted with self-satisfaction nor logically acquiesced in if one also accepts the conviction that one's nature is rational, and to be rational is to be one's coherent and unified, whole self. Otherwise expressed, it would be self-contradictory for a rational being to accept as final a disparity between his felt self, his cultural, and his complete self. Thus, he accepts the rational self as his whole self. If on the contrary one rejects the interpretation of oneself as rational — and seriously undertakes to put this decision into practice — one may be in no position to judge the outcome of this practice. Such a person will never be in a position to know whether his choice of means achieved the end which he posited.

In consequence of some such reflection as this the individual is persuaded to risk a definition of himself from which those obligations follow that dignify him as a rational person. He thus identifies himself as having awareness of or insight into the universal and real self, an insight which can be rendered ever more and more perspicious.

Thus, obligations are recognized and exist for us because they follow from our acknowledgment of our rational nature. Guilt is the consequence of the denial of that nature, a denial all too easily effected by identifying rationality itself with some successful technique or prestigious role. In any event, the identification of the self with rationality (*logos*) expresses a long standing intuition of the self. This important point is susceptible of formulation: the person (A) is obligated to his whole and rational self (B) to perform C whenever the omission of C would prevent his recognizing the disparity between the feeling and the cultural self or between the cultural and the whole self. It follows that the primary duty of the person who acknowledges obligations is to know or to endeavor to know the truth about the complete self. Such is his wisdom. This contemplative ideal, or something like it, has a claim to being the guide which has led Western man out of the caves of barbarism. The Western world accepts it; only this world holds that individual man is powerless rightly to direct himself by his own efforts. He requires, therefore, the assistance of God, of Genius, of organized and corporate man, of future science, or of super-technology.

Even with this assistance, the C in our formula will doubtless remain obscure, such is the consequence of the limited rationality of the creature owning to the ontological defect. Failing, thus, by his very nature to see or to be himself or to see just how he is not himself, this defective creature may be said to be ontologically guilty. This is the ironic aspect of the faith to which the West has clung. Perhaps this irony can be read into the Christian rejection of secular wisdom. Man can aspire only to the love of wisdom and to the recognition that he does not and cannot possess it; although, it may be given to him.

## §16. *The Ambiguity of the Whole Self: The Questioning Self*

This view of the defect in man's being, the inverse of the conclusion to its dramatic richness reached in the preceding chapter, prompts a final characterization of the self's primary obligation. It is not enough to observe merely that the self is the being which can manifest to itself its own being and that of other things, nor to note that its self-interpretation, its pursuit of completeness, proceeds by way of a non-objective dialectic. Here we add that the self is unique in possessing an obligation to engage in the pursuit of completeness. Obligations were specified as a set of valuations to which the person is committed both affectively and in principle; they form an indispensible component of the self. Yet the cultural standards which pass judgment on obligations cannot be supposed really to define the self; although, that which sees, accepts, and imposes these standards may be the self. Neither are the feelings and impulses which are judged, identical with the whole self. An argument was adduced which concluded that the classic humanistic definition of man is a very persuasive definition; the obligations of men are to be recognized as rational obligations. They arise from human being itself and his world and indicate the primary directedness of the human power. They are accepted by the rational self and imposed because they are rational or human. By disciplining his feeling to accord with this rational acceptance of obligations, the existent self can envisage and move toward its complete self. Thus, we concluded, if X is a man, X is obligated to himself to become his rational or human self. Guided by this conviction, a man can expect to become free of guilt.

Hardly, however, is this conclusion reached than its cogency fades. For if history has made any one thing evident, this is that rationality, no doubt a clear and definite concept to the Classical Greek philosophers, has become an ambiguous notion.[5] Men of almost all times, places, allegiances, and varieties of commitments are insulted when they are called irrational. Is the rationality which they severally desire equally

various? Is it possible now to define rationality in an unam-
biguous and commonly acceptable manner? If not, if the nature
of reason itself has become questionable, then the man who
regards himself as the rational being must also regard his being
as questionable. Such is, I believe, the proper sense in which
ontologically defective man may regard himself as rational: he
is in pursuit of the truth about himself and his rationality. A
man exists under the obligation not merely to examine his life
but to lead a life of constant and fundamental self-questioning.
This obligation should constitute a primary motive for action
and a basic element in character structure. In effect this view is
inherent in our earlier statement about the self (p. 42): that it is
identical with its possibilities, and that these possibilities in
number and nature are initially indefinite.

Here I offer testimony to varying views of the nature of reason
by citing briefly two different ways of understanding this nature
which have emerged in the West. The older and more complex
notion is derivative from the ancient conviction that man's
subjectivity, which can be disciplined to rationality, is dual,
being characterized by poetic powers of intuition and insight as
well as by ratiocinative capacities. [6] No doubt both of these kinds
of functions need to be developed to their full and cooperative
maturity. But the difficulties in coming to an understanding of
the poetic or imaginative capacities are certainly enormous
today, and yet a long and persistent tradition has held that they
exist and are, along with one's ratiocinative power, one of the
indispensable means for working out one's fate. Indeed, insight
into the perfect life possible for humanity was claimed for the
poetic dimension of reason and expressed by Plato in poetry
about the chariot of the soul (*Phaedrus* 248) attempting but
failing to follow in the procession of the gods. We have noted
that Christianity developed this sense of man's incompetence
into a conviction of the incommensurability of any current self,
however rational, with the complete realization of human
powers. It estimated equally highly an obedient receptiveness to
inspiration and a faithful yielding to the Spirit. Outside of the
religious context, expressions of a similar kind have been used to

describe not only the poetic mind but even the originative scientific intellect. They apply with peculiar appropriateness to the man governed by the obligation to himself to utilize all his powers in the quest after his complete self.

On the other hand, a simpler and, in much of contemporary opinion, the more prominent conviction about the nature of rationality eliminates the ancient receptive and poetic psyche and tends to regard the discursive and ratiocinative intellect and its aids as exhaustive not only of the mind but of the self. Man, according to this conviction, ought to pursue clear and precise concepts about nature, for this knowledge is the power which may be directed toward making him the master of his environment. And this mastery, which only corporate man can approach, is the indispensable means for the salvation of the individual man. The scientist-technician, member of a team of experts, is the modern hero whose dealings with fate are regarded as exemplary. Indeed, it may even be that the scientist-hero is finally identified with fate. The modern self admits, of course, to defects in the individual, but he places his faith in the collective attack upon technological problems and expects, through the endless progress of reason and the scientific method, to enjoy ever increasing prosperity and continuously lengthening life, even though actual achievement of this desideratum may be deferred to the indefinite future. Individual guilt does not loom large on this interpretation. It is regarded as something like a remediable error in technique or as a curable illness. The well adjusted man is free of it.

We may look upon these two conjectures about the nature of the rational self as two possible decisions. The first and more complex one appears to lead by way of an internal dialogue toward identifying the person primarily through self-questioning and through cultivating recognitions of his hybritic self-misidentification. The second would ignore this possibility and would turn rational activity toward problems related to natural objects or toward what Kant calls the empirical self (the self as object) with the expectation of discovering positive knowledge about, and finally power over them. The former is the tragic way, the latter the way of scientific reason and technological progress. Very probably history has not yet drawn out all the consequences for the self and its future implicit in either of these views of reason (cf. Chapter VII). No doubt disaster would follow upon the choice of either of these two kinds of dialectic to the exclusion of the other. Hence, individual persons are rightly advised to recognize the absolute and decisive nature of this crossroads, to weigh the alternatives in the light of insight into their possibilities, and into their valuations and obligations, and to arrive at a rational decision — although such a suggestion is not innocent of irony. Perhaps this irony is most surely retained if the individual person learns to think of himself as engaged upon both of these ways. Then he must surely recognize himself as a being primarily in search of his identity, or at least as a being whose insight into his identity can and should continually be deepened in the interest of harmonizing the opposing interests and impulses of his complex possibilities. The final ignorance to which Socrates and others have laid claim expresses this kind of questing or wondering reason and provides the clue to the dialectic by which this kind of reason is realized. In an ideal humanistic society, no doubt, all persons and institutions would serve this primary end, whatever subordinate ends they might also seek. [7]

Thus, in summary, the phenomenon of obligation is the phenomenon of being bound to certain kinds of valuations and actions, even though others might appear to be possible and attractive. The source of this boundedness, the reason for obligation, lies in the self which imposes obligations, especially the primary obligation to seek and to be guided by the truth about oneself. This self has appeared in several interpretations: the ancient self, seeking identity with the felt father-authority who imposes judgment and assigns guilt; the cultural or hypothetical self which one normally aspires to become; the whole self which one presumably always is (in possibility) and which seek rationally to unify the two former selves; finally, the questing self which recognizes that any identity or unity of the self which may be accepted is in truth questionable, and that only by way of this questing (i.e., the way of the non-objective dialectic) can one enter in on that change which is the life and renewal of the self. St. Augustine's remark sums it up: *"Quaestio mihi factus sum."* Thus, our guiding declaration (p. 32) is justified: it expresses the primary human obligation.

We hesitate to conclude upon such a note, however, for the view that man is forever in quest of his nature or essence, or that his reason is properly a questioning, is tantamount to a denial that man has an essence which can be conceptually defined and used as a standard for evaluating other views of man or for evaluating proposals concerning his final good or his appropriate way of life. Are we, then, fixed in a cultural relativism, condemned to be without a standard for evaluating the human good? Such a predicament would indeed be embarrassing for a philosophy which set out to take the human measure of technological culture. But in fact there may be a way around this predicament. The next two chapters, one on truth and one on praxis, will seek to extricate us or at least to legitimize and to utilize the predicament.

CHAPTER IV

# ON TRUTH, SUBJECTIVITY, AND SELF

## §17. *The Task for This Chapter*

The end of the last chapter pointed to a primary obligation of the self continually to question itself in the endeavor after self-knowledge. That chapter together with the preceding one also indicate that the self is not a definite object possessing a fixed nature or essence. Thus, the self may not remain constant, like the relation between the tuned strings of a musical instrument or like the interrelated functions needing to be performed in and for a stable society. As identified with its possibilities, the self cannot be got into a laboratory or defined by means of investigative or of mathematical techniques as if it were an object. Nor can the nature of the existing or actual self be determined by means of scientific techniques, since this self has a non-objective aspect. Moreover, the very process of questioning it tends to elicit new possibilities and so to change it. One might better say that its laboratory is the theater, the arena of human change. If, then, the self is obligated to pursue the truth about itself, and since the self does not possess a definite and permanent non-temporal nature open to usual methods of investigation, its relation to this truth cannot be like the relation of data to a theory which explains them. What, then, is the relation of the self to truth about itself? By what route or routes does the self experience itself, discover, and come into possession of truth about itself?

According to our opening ontological statement (§4), the being of the self is power directed toward manifesting its own being and the being of other things. A man interprets both himself and other things. As it may also be expressed, a man can, within limits, vary the directedness of his given powers; that is, he can direct himself. To discover the appropriate directedness, that which fully elicits and manifests his power, is man's first problem and his primary obligation. According to Western tradition, the possibility—even though always imperfectly realized—of seeing or discovering this truth is an essential possibility of human being, that without which he would not be human, would not be self-experiencing and self-directing, and could not, therefore, discharge his primary obligation. This power of discovering or seeing now claims our attention. We shall have to ask: what is the general character of that which is seen? What in man is that which sees? How is this seeing related to self-interpretation?

In this essay the name "subjectivity" is given to a man's power of standing consideringly outside of himself, as it were, or of separating his powers from their directedness and modifying or redirecting them. Thus, subjectivity is the possibility of seeing or discovering and finally of using truth. It is that by means of which one is receptive of presentations and interprets them. By means of it a reflective and human experience becomes possible. The notion of subjectivity is peculiar in an important respect. Like the Kantian imagination (*Einbildungskraft*) it is partly an ontological, partly a non-ontological and cultural (even a psychological) notion. In its ontological function, subjectivity embodies the world-structure of an epoch and thus it opens out a definite and limited type of perspective upon beings. In its other function, it is that part of the human power which can be disciplined within this perspective to the search after truth in a manner broadly limited by a culture and developed by the individual. This function and certain of its techniques will be the main topic of this chapter.

Prominent in the structure of subjectivity is the self. The self is felt to be the center of subjectivity in the sense that it owns it; that is, all possible or actual experiences are referred to the self, some being more closely identified with it than others. The self is, at least in possibility, in touch with all parts or aspects of his subjectivity and ideally can use them to accomplish its goals. In fact, of course, a person falls far short of this wholeness and must, therefore, use his powers and whatever additional aid he can muster to discover what this wholeness is and how to gain it. Chapter II introduced the view that the self comes into relation with itself and approaches its wholeness by way of a non-objective and dialectical relation with the truth about itself. Chapter III noted several stages in one possible development of the dialectic of the obligation to pursue this truth. In addition, by way of an objective dialectic the self comes into possession of truths about nature. Some aspects of this process will be considered in Chapters VI and VII. Through a kind of combination of these two dialectics, the self develops its relation to other persons (we may call this the dialectic of the other). In this chapter we shall be occupied with subjectivity and its operation in the non-objective dialectic, whereby it seeks out the truth about itself and seeks to acquire the power of self-direction. It will touch upon the other two forms of dialectic only incidentally. Section 18 will provide a relatively abstract account of truth as ordinarily understood together with emphasis upon a subjective factor usually neglected. The next section (19) will describe some of the relevant properties of subjectivity, the power of seeing. The last section (20) will then offer an account of this subjectivity functioning through the non-objective dialectic to discover and to use the concrete truth about the self.

Our attention will be directed first upon truth which is to be understood, then upon the understanding of it, and finally upon the one who understands. We hope, thus, to shed some light upon the task of pursuing the truth about the self. Ironically the drift of this chapter will be further to verify that we cannot discover a fixed human essence clearly valid at all times and in all places, and standing in judgment over the varieties of its cultural expressions. This chapter, nevertheless, together with the next, will discover certain properties and events of human life which do provide sufficiently for this need.

## §18. *Empirical Truth and its Generalization*

The term "subjectivity" is used to name the characteristically human possibilities of the self. Disciplined subjectivity, or reason, has traditionally been identified with the power of systematically seeing, naming, and judging the relations, similarities, and differences among things, among the ratios in things, and among ratios. Reason, then, refers to those powers of subjectivity which are disciplined to the search after truth. It has commonly been thought to find a particularly proper and fruitful use in the pursuit of insight into the self and into the life appropriate for man to lead. Since the renaissance, however, as we noted in section 16, this rather general view of reason has been narrowed. It is often identified only or primarily as the power of judging the ratios in numbers or mathematical entities and the equality or inequality of their ratios with the defined objects of the sciences. Cartesian man is man self-limited to performing primarily these narrowly rational functions. When Cartesian man judges or is shown, by the mediation of quantification and accurate measurement, that a mathematical statement is regularly reflected in objects, then he holds that he knows a truth about these objects. More generally, we believe that we know the truth about a collection of objects or data when they are explained. And a collection of data is explained when it is shown to be an instantiation of a set of precise theoretical or abstract propositions, or statements of propositions. Then the data correspond to the abstract statements or theoretical model. Or again, the concept is the analogue of the data and, hence, is true of them.

Now it might seem that if the concept, image, or belief is

explicitly shown to be an analogue of the data, then it would be all but tautologous to observe that the concept, image or belief is true of the data. Yet the apparent tautology fails to reveal an element without which truth is quite unintelligible. This element, present in the older sense of reason, is usually missing in the modern sense.

Consider first a simple illustration. The truth of my eye measurement of the equality in length of the two tables in this room could be perceptually verified by placing them alongside each other so that one could be seen truly to measure the length of the other exactly once. Is this correspondence of the two tables with each other all that is normally meant by observing that one is truly equal to the other? Hardly. A more complex illustration will help to elicit the omitted elements. Suppose that one of the two were in another building. Then the vague impression of their equality in length could not conveniently be verified except by resort to conceptual mediation. To effect this mediation, a unit length may be defined and operations for its use specified. Then the unit length might be shown operationally to be related to one whole table length as the number one is to (say) the number seven. Then the same isomorphism between lengths and numbers could also be shown to hold between the other table and the ratio $1:7$. Thus, the ratio $1:7$, being a concept and applicable in both buildings, mediates between my perceptions of the two tables and enables me to judge truly that both correspond to each other, both being seven feet long. An element present in the second illustration and seemingly absent in the first is the concept of a specific quantity, "seven feet long" (or the ratio $1:7$).

There are two points to note. The first is that conceptual operations are not really absent in the first illustration, but they are not obvious. By placing the tables side by side either of them was tacitly used as the unit of measure. And it measured the other table exactly once. Thus a concept, signified here by the ratio, $1:1$, applied in both instances, just as in the more complex illustration. The second point to notice is that the concept, a subjective entity, performs a function which the unit of measure and the operations of measurement do not perform. Probably the second illustration best elicits this function. Consider that if X measures A and X measures B, then A measures B or they are equal in respect to the measure. Here it is evident that one must "see" the transitivity of the correspondence relation in a sense different from seeing the two tables placed side by side. At least the function of the concept as a "seeing" is more obvious when the tables compared are in different rooms. The two functions which the concept is performing can be expressed in this way: the correspondence of the two tables is true in one sense as that about which we truly judge, but the conceived correspondence is also true in another sense, as that by means of which we judge. Thus understood, the concept is endowed with a new, non-logical and non-objective function: that of judging truth. Thus, the concepts (1: 1 or 1: 7 in our illustrations) to which the tables correspond, is also that by which we become aware of the correspondence.

This dual function of the concept (image) may be expressed generally. When we judge a concept (or image) to be true we mean (a) that it corresponds with the relevant facts and (b) that this correspondence is seen to hold. But the mere assertion that something (a concept, image, interpretative principle) corresponds with something else (e.g., a table) is not identical with the assertion that such a correspondence holds *and* is seen to hold. Perhaps the former, the correspondence, is the necessary condition for truth, but the latter, the seeing of the correspondence, is its sufficient condition. The two conditions, however, are so intimately linked that the seeing is often either overlooked or else identified with the corresponding. The fact of the conceptual or imagined correspondence is regarded as identical with knowing the truth. This identification, which in effect awards a privileged status to the mental correspondent, confuses factors which are discriminable and must be discriminated if the sense in which man is interested in the truth about himself and is obligated to pursue it are to be understood.

Sometimes the seeing function of a measuring concept is concealed by being passed along to other concepts or operations. For instance, it may be passed on to the notion of verification. The number, X, which I am now considering, may correspond exactly with the number Y of inhabitants of the United States in 1984. But I do not see or know that it does. In order to conclude that it truly does, this correspondence would have to be verified. Verification is usually described in terms of rules for collecting and using evidence intended to confirm or disconfirm an asserted correspondence. But again, these rules cannot be applied blindly. They, too, must be seen and understood in order to be applied. Thus, they too perform a dual function: they direct one in the collecting and using of evidence; and they presuppose that evidence is seen to be evidence. Evidently this seeing is presupposed by the theory of evidence as well as by the theory of truth. But the latter theory is the prior one, for the evidence which verifies is expressed in propositions or concepts which must be seen to hold or to correspond with the verifying data. These propositions or concepts also perform a dual function: they express or set forth the relevant correspondence, and if true, they are commonly identified with the seeing of their truth.

This silent identification of the concept or expression of correspondences with seeing truth may be placed in perspective by generalizing the notion of empirical truth. For empirical truth, as illustrated in empirical explanations, is only one species of a generic truth relation. This relation has been described at various times in the history of philosophy. For example, Plato, Leibniz, Kant, Whitehead and others offer similar descriptions of it. The general description says that whenever the elements of two classes can be shown to correspond with each other in respect of some relation, then we may say that a truth relation holds between them. Then either class may be said to be "true of" the other. Thus as Descartes showed, algebra is true of Euclidean geometry (or vice versa), and as Newton showed, mechanics is true of large physical bodies. A convenient expression of this truth relation is the general form of analogy, $A:B = X:Y$. Here the symbol "$:$" stands for a relation; the equality sign says that the relation is the same or similar in the two instances. Then the class A: B corresponds to the class X: Y (i.e., A corresponds to X, and B corresponds to Y). As stated thus, no mention is made of the act termed "seeing." But without this activity a correspondence is not brought into the realm of human experience; much less is it made public.

The truth relation plus the subjective operation of "seeing" or intending were tacitly utilized in section 4, where perception was described as interpretation of the given or presented by discovering some degree of similarity between the sensed presentation and the standard representation or concepts which serve as the principles of interpretation. Chapter II utilized a similar schema, only that which was to be interpreted was one's felt possibilities, and the interpretative principles were either the standard roles provided by a society or principles more personal and peculiar to each, discovered by something like poetic insight. It is not difficult to show that analogies or metaphors in poetry use this same relation in a way not generically different from its use in the sciences (§33). The term "insight" is aptly used in this regard, for it suggests seeing in between two classes, e.g., X (A,B,C . . .) and Y (x,y,z . . . ) in order to take note of the correspondence or its absence among the elements of the classes and of the similarity between their ordering relations (X and Y). The recognition of some such correspondence among elements of different classes is surely a basic recognition.[1] But this recognition is something over and above the correspondence itself; this recognition is just that something added by man and has its being in his subjectivity.

The advantage of expressing the truth relation in a generalized form arises from the circumstance that the correspondence between object and concept is seen to be only one species of the general correspondence or truth relation, though not unlike the correspondence of presentation to interpretative principle, of fact to fact, of concept to concept. Thus, it becomes more difficult to accept the kind of cognitive privilege with which the mere correspondence of object to concept or concept to fact has been endowed. The correspondence of fact to concept is not necessarily possession of truth about that fact; the showing and seeing of this relation is an essential factor.

The point which I wish to convey is one which preoccupied Plato. He was struck by the strange circumstance that one and the same relation might be seen to hold both between A and B and between X and Y. The presence of the same relation in different instances has elicited both wonder and confusion throughout the history of thought. That twoness should be fully present in several different groups of (say) tables without in any sense being used up or diminished has so amazed some philosophers that they find in such entities a separate and special sort of existence (or subsistence). One of these philosophers was the wise old Chinaman who pointed out to his disciple that the cow and the calf grazing in the field were not just two things, as the simple-minded disciple supposed, but three: the cow, the calf, and the both of them together. Plato, though, was more conservative in holding that their twoness is not a third thing but just a participation in duality which is precisely not two ones which make another, a "two," but one pair. This pairing which constitutes a perceived or understood truth can not possibly be another thing like the objects paired. It is not even a thing idealized. We may speak of the pairing as an activity, a grasping, a seeing, or an illumination. But it is essential to recall that the action of seeing or illuminating a correspondence is not another such correspondence. (To suppose it to be another such correspondence is to land oneself in that philosophical bog, the infinite regress.)

It is not easy, however, to know how this all-important seeing is to be further identified. Plato offered the so-called myth of recollection to indicate the direction in which this aspect of his theory of truth was to be pursued. The myth says that the mind already sees or recollects duality and thus can recognize it again in any collection of two objects. The mind is that sort of organ which, having already had experience of the unit, and of the dyad, can now recognize the unit object and then count objects and stop when it recognizes the number two. The concept of two, like any concept re-presents an original seeing, a seeing, moreover, which can be extended, as in this illustration, to illuminating present events and objects. What is involved in possessing this standard unit and this operation of counting that true propositions should issue from their use? The assignment of an arbitrary conceptual element to an object and the operation of counting are not obvious and self-explanatory processes. Yet somehow we do apprehend abstract and concrete elements, we do pair them off or see them together, we do retain past groupings in mind, and we do recognize when a definite collecting of this sort has been accomplished. This is a sophisticated and complex action; Kant described it as a three-fold subjective synthesis. Kant, though, knew that this synthesizing was not a mere correspondence nor was it an independent, self-sufficing activity. Rather, as he points out, the "I think" must accompany, and as it were, unify any such operation. This Kantian transcendental unifying is a reference to the one who sees.[2] His is the seeing which the correspondence theory of truth, as it is often stated, carelessly slurs over; this slurring has the effect of endowing one selected type of correspondence (the correspondence with a concept) with a privilege. The result, as I have tried to show, is an ambiguous theory of truth. In this ambiguous theory, "truth" sometimes refers to the truth relation, sometimes to seeing that a truth relation holds.

Both Plato and Kant recognize that the mind always already sees, and that the source of truth lies in the seeing, in the entertaining of a correspondence, not in the correspondence entertained. I think Plato and Kant were on the right track. Correspondence or truth relations (however established) are what we know when we know truth. An indispensable power of the self is this power to see, to understand, to intend, without which it could not, for example, become aware of its obligations. This power is termed subjectivity.

## §19. *Some Properties of Subjectivity*

We turn now to subjectivity which sees or discovers truths, in particular truths about the self. Subjectivity, however, is difficult to see. I have noted (§11) that a poetic use of language has been more successful in evoking the self than either a scientific or commonsense use. It is not surprising that attempts in the past to understand the nature of subjectivity have made free use of metaphors, metaphors of light as well as of seeing. We shall take both metaphors into account. If these metaphors succeed in directing the intellectual vision, this section itself will offer an illustration of a typical operation of subjectivity.

One difficulty in perceiving subjectivity arises from its lack of object properties; in consequence of this lack, it bears an unusual relation to time and space. Notice that subjectivity, or the person associated with it, lives and develops in dramatic time.[3] In a person's experience of himself, his decision concerning his identity or function made in the past lays out a path in time and comes toward him, as it were, from the future, directing and determining his behavior in the present, at least until the decision proves invalid. Other kinds of experienced time can be shown to fall within, or to be abstractions from, this concrete, finite, dramatic time. Similarly, the space of subjectivity is unique and concrete. No doubt subjectivity can be located in space; normally we are accustomed to identify subjectivity with the localizable body of an individual person. But that it is "spread out" over any part of his body can be both affirmed and denied (cf. p. 35 f.). Upon consideration, its location at the body's location can also be concluded to be ambiguous. A Parisian may always be in Paris in a perfectly intelligible sense, even when he is seated in a cafe in Detroit. We attempt to avoid the ambiguity by saying that he is subjectively in Paris although objectively in Detroit. But the relation of subjectivity to its location is not so easily clarified. For the Parisian may be quite subjectively effective in Detroit. Evidently the concepts useful for specifying the date and location of an

object like a cafe are not adequate for describing the time and
place of the subjectivity which identifies, locates, and dates the
cafe. Doubtless, therefore, it will be appropriate to lay aside the
assumption that subjectivity is like an object and attempt to
approach it directly by observing the way it is in fact experienced.

Subjectivity, we have observed, is experienced as a kind of
seeing or as analogous to seeing. Consider again the seeing that
a correspondence of some kind holds. Before being discovered,
a possible correspondence or truth is neither true nor false but
unseen. Now discovered truth does not just happen; nor is it
presented to one as a gift. Men, as C.S. Peirce has noted, have
repeatedly moved in a most aptly chosen direction toward a
novel truth already correctly conjectured. Similarly Henri
Poincare remarked that men, at least some men, have a
"subliminal self" which with a strange accuracy selects concepts
which are true of a given class or situation. And thus, to the
question: "How in principle does one come into the possession
of truth or direct oneself toward finding an undiscovered
truth?" we respond that in some curious sense, we are already
forearmed. We are led again to the suggestion drawn in the
preceding section from Plato and Kant.[4]

The present inquiry is not an idealistic nor a Kantian study of
the conditions of knowledge. Its aspirations are less lofty. We
begin with what anyone might observe. In this instance, we
observe that we can attend either to this thing or to that, or we
can attend to ourselves, but not to everything at once. Obviously
attention and inquiry cannot be turned in all directions at the
same time. How is the chosen direction selected? The world one
inhabits first directs human powers into certain broad channels
and away from others. Then this directedness is modified and
interpreted by culture. In our intellectualist world, one does not
attend to a thing solely in its individuality, at least not
habitually. One attends to a thing as similar to or in some way
corresponding to some other thing. Also, the Western world
inclines to give special attention to objects that can symbolize
aspiration and infinite reach. Thus, when in Saint Louis I

cannot but be drawn to the contemplation of the Saarinean Arch. To me, a man of Western culture, arches symbolize a movement upward and outward. As an American, I find — I am surely meant to find — my imaginative gaze impelled endlessly beyond by a spontaneous interpretation of the shear size of this Arch. On a less symbolic plane, I can respond to the Saarinean Arch merely because I already know, though in a generalized way, what an arch is like; I already know that this Arch would be similar to the standard arch-representation with which schooling in a Western culture has equipped my memory. This instance is typical; I risk a generalization. It is easy to verify in common experience that the correspondence of one aspect of experience or of the world to another is normally conjectured or vaguely intended before being searched out in detail. Only so would the mind direct itself toward seeing one truth rather than another.

By a similar operation, linking expressly together that which belongs together, the mind purposefully assembles the elements of language; it intends a meaning before expressing it; it can finish a sentence it has begun. If the mind were not guided by some conjectured grasp of its meaning or goal, some pre-apprehension of the truth it seeks to express, then it would proceed in no particular direction to make accidental associations. Human activity, however, is not random. Indeed, then, the mind must be equipped with certain types of pre-apprehensions or illuminations by the very fact of its being in a world and having a culture, however inexplicit the basic types of pre-apprehensions may be. What is uttered, thus, develops out of a basis in the prelinguistic; similarly, cognition develops out of a ground in the pre-cognitive; the seen grows from roots in the imprecisely visualized, and the humanly actual is founded in a subjectivity which is never without structure. By means of this structured subjectivity the being possessing *logos* may come into possession of a world.

One may find an illustration of this obscure pre-cognitive apprehension which guides perception and inquiry in the experience of searching for a name nor remembered. A somewhat

different and more dramatic illustration comes from the poet's
effort to recollect just the right image to express the still unborn
thought. By what standard does the poet reject the many
competing images which come to mind? In copying a model of
some sort, the standard guiding the copyist is obvious enough.
But in any sort of creative work or in novel perception, this
standard is precisely what is not known and is not seen; it only
gradually becomes evident, usually after much trial and error.
And yet it is grasped even before its expression with sufficient
precision to enable the distinction between an acceptable and
an unacceptable trial to be effectively made. Here the function
of subjectivity of containing, as it were, or of somehow pointing
to that which seems to be unknown or unpossessed, is even more
evident. I always find myself already obscurely seeing that
which I seek. A similar experience of the not-as-yet-explicitly-
known guiding to the confident possession of truth is characteristic of the one who makes and expresses discoveries in a
science. The history of science is replete with illustrations of the
way to new knowledge being opened up by way of analogy with
the old, even though at times the analogy appears to be remote.
And how does the scientist use this way to new discoveries? The
comparison, popular today, of a computer scanning a large and
complex collection of memory cells is not illuminating. For the
exact cell according to this comparison must already be present
in the memory bank; the only problem lies in selecting or
remembering it. The creative discovery, however, is more like
adding a new circuit to the computer which enables the
machine to perform new tasks which the inventor had already
vaguely visualized. But there is no algorithm for making such a
discovery.

Subjectivity, then, is experienced as the power of seeing or of
recognition; at the same time it is the power of seeing ahead
which comes into play to link the earlier stage of a purposive act
with its later stages, no matter how brief or lengthy the act.
Likewise subjectivity is experienced as a power of originative
seeing, or meaning-giving, which comes into evidence in a

creative accomplishment. Perhaps it comes most clearly into evidence when the self, in the process of change, seeks to direct its current self to become the complete self. Thus, as the predisposition to truth, subjectivity is the power to see or to attend, to attend prior to expression or cognition, to originate meaning, particularly its own meaning. "Subjectivity" as the name for these related elements of a self's possibilities refers to that without which a self would surely not be human, would not be self-interpreting, would not be a being having *logos*.

That these properties belong to subjectivity seems to be evident, still it is not clear that accumulating a list of these and other presently accessible properties will express with precision just what subjectivity is.⁵ Moreover, the essential property, a seeing or a seeing ahead, and especially an originative seeing, are remarkably elusive; are we not in danger of finding that anything which we do say about this aspect of subjectivity cannot be said to be verifiable? Admission that subjectivity is not an object and, therefore, cannot be measured, hence, does not figure in empirical theory, seems to invite this danger. These remarks, however, on truth and on the subjectivity which complements the correspondence theory of truth, are prompted by common experience and are of the nature of philosophical theory. A philosophical theory of truth, however, is not analogous to a scientific explanation which is said to be true because it is verified; rather, a philosophical theory of truth seeks to become aware of the seeing which selects one account of truth, or of verification, rather than another. The necessary element in a theorizing of this kind is just that which can bring a type of correspondence or truth-relation out of the limbo of the unseen in which it cannot yet be said to be either true or false. We might liken subjectivity to that context within which a computer was invented, has a meaning, and can be used. However, in the philosophical tradition, this seeing function has more frequently been compared to light, a "natural" or given light, which lights up certain relations so they can be seen to hold. This light is the subjectivity of which a theory of truth should make us aware. Use of this second analogy is still desirable since it calls attention to factors which otherwise might be ignored.

The conventional objection to the comparison of seeing objects to illuminating a scene is offered on the ground that it is metaphorical, hence derivative and unclear. This facile rejection may seriously be questioned. How do we recognize a source of light, say a street light lighting up a roadway? No doubt because we have seen it before. It is a matter of common experience, part of our usual dictionary of images and concepts. But how did we recognize it before? Possibly because a teacher taught us how to use the proper descriptive phrase in certain instances. How, though, do we know that the phrase applies properly in other instances? Because, it will be surely said, the expected consequences occurred when the phrase was used. But again, how are the expected consequences recognized as such? Because there is a correspondence between expectation (concept or image) and actual perception? And here, we are back with the correspondence relation which we wished to illuminate. There is no egress, I think, from this labyrinth unless somewhere there is a seeing, a recognition that although one light lighting up a scene is like enough to another, still the *seeing* of this resemblance is a unique operation, without which there is no naming or describing at all nor any recognition of objects, images, or correspondencies. We rightly think of this seeing as an original illumination of objects without which there literally is no evident thing, nothing discovered. The image of the street light is only one of those things which is illuminated by this original illumination. Perhaps the myth of Prometheus, the light-bringer, alludes to the basic and originative character of this initial illumination. [6] Supplementing the metaphor of seeing by this metaphor of illuminating will help to bring other aspects of subjectivity to attention.

The fully elaborated analogy to light would have to make reference to the source of light, to the illumination, and to the objects illuminated. Here we have the self—the one who sees— as source, subjectivity as the illumination, and truths or correspondences as that which are lit up and seen. The analogy used thus is an effort to describe the way in which one sees one's own seeing. The analogy suggests that as an illumination has a definite source, so subjectivity must issue from a center, an origin, which may be located or in some way specified. Again, we should recall that the source of light is neither the illumination nor, in every sense, one of the objects illuminated. The source is itself seen and is a condition for the illuminated object and for the illumination being different from the surrounding darkness. Also we should remember that more than one kind of illumination is possible, and that the kind of illumination used is a controlling factor in determining what comes to be visible. And finally, we recall that the light once lit may be redirected in various ways and that it may be extinguished.

It will be helpful, before utilizing this analogy for our purposes, to replace it in the context in which it came to its classic expression. There its complexity appears. For this light is not ordinary light but a *lumen naturale* which is also a seeing and which enters in a certain sense into that which is seen. Plato's views on vision provide the fuller context. For in effect Plato held that like knows like; [7] hence, the illumination from the sun had to be met by something like light emanating from the eye of one who sees. Hence, he held, a "stream" must issue from the eye when it responds to the sun's light. Only when the two streams mingle, marry, and so correspond does perception come to birth. Thus, illumination or subjectivity is amplified in a sexual metaphor which combines seeing (from the eye) and light (from the sun). In sum, the power of the "eye of the mind" has to be inseminated with the light from the sun of the Good before insight or intellectual vision can occur.[8] As it might be imagined, the original separation of Ouranos and Gaea (cf. §4A) is reenacted in every instance of illumination or seeing a truth. The seeing of that which is illuminated is a seeing "between" the two classes of elements which correspond; it elicits their correspondence. The "stream" from the eye or the readiness of the intellect to "see," must respond to the solar illumination; and the provocation of the good, before vision of either kind can occur. Thus, a possibility issuing from the self may through action become involved in the world and eventuate in an insight (cf. §22). The point is that any seeing of a truth presupposes an originative (presentational, interpretative) vision. It presupposes first, a presenting of the beings which appear to the receptive sense and, second, an interpretation or a meaning-giving of *logoi* to the beings which are thus presented or encountered. Subjectivity is a power for receiving and uniting with a something presented and interpretable. It brings the perceptible to perception and recognizes the intelligible. In consequence it sees objects which are there to be seen and sees the correspondences which are perceptible among them and between them and concepts.

The analogy to light also points to the limitations native to subjectivity, for it suggests that subjectivity like an illumination is not merely a characterless power of "seeing"; rather, in any instance it possesses a definite structure. We can respond to things only as subjected to a limited illumination. One way in which this light is limited is in being definitely directed. Directed illumination allows certain kinds of correspondences to emerge into awareness while obscuring others. The terms world-structure and world have been used in this connection. A world is determined, as it were, by the angle of illumination (perspective) which is favorable only to certain appearances. Also a given subjectivity is, so to speak, able to receive illumination only of certain colors. It can grasp correspondences only in certain (reflective) contexts and in certain contexts of contexts (cultures). It is never sensitive to an omni-directional illumination or to the whole spectrum.

Likewise, the illumination is limited by varying as if in intensity; that is, the subjective illumination always possesses a distinctive character which we describe in terms such as satisfaction, curiosity, anxiety, matter-of-factness, objectivity, and the like. Perception cannot divest itself of some such modification, of some *way* of perceiving, or of the clarities or distortions which are thereby introduced.

Just this limited power of interpretation, this characteristic kind of rationality, defines one's predisposition to one kind of life rather than to another. This relationship to one's kind of life is sometimes expressed as the correlativity of man to the world in which he resides at any epoch. As the stream from the eye must mingle with that from the sun in order that vision should occur, so a subjective readiness to live in such and such a culture must encounter the circumstances which elicit that kind of interpretation before human life of a particular kind can occur.

In noting that human subjectivity is the predisposition for certain types or categories of activity, I mean that it is sensitive at any time to distinctive and limited groups of problems and types of solutions. These define the kind of difficulties a person gets into as well as gets out of. They specify the obligations by which a person feels himself bound and the roles which seem to him the reasonable ones to play. They reveal his world and his culture. The meaning and value of events appearing to any one exhibit concretely what he has done to date with his subjectivity. Conversely, if one could read subjectivity no doubt one could read in it one's world, one's culture, and one's own habits of seeing.

The metaphors of sight and light have made it evident that subjectivity is always directional, culturally modified, and specified by some individual *way* of awareness. Subjectivity sees ahead, and at times sees originatively. Thus, it animates language and renders possible relatively independent interpretations. In consequence, two objects or classes, say X and Y, do not merely correspond; they may be seen systematically to correspond in a characteristic and concrete way from the point of perspection which a given world and culture permit.

We have in this section extracted from the metaphors of seeing and illuminating something of what can be said about the subjectivity in which experience originates. And now the question would be reasonable: can we not have done with metaphors and speak literally about subjectivity? The question, however, exacerbates the predicament: the being who possesses *logos* may not be literally describable by means of his *logoi*.

Sometimes his characteristic power, his subjectivity, is said to be the power of giving meanings. Recognized objects, true correspondences, are among the meanings (*logoi*) which are given. Recasting the accounts of subjectivity as seeing and as illuminating in the language of meaning-giving, however, is a doubtful clarification. The movement is only from one metaphor to another. Subjectivity is not literally a giver who gives the pittance of meaning in due proportion to whatever deserves it. Also meaning either as act or as concept is no better understood than subjectivity itself. These several metaphors, nevertheless — seeing, illuminating, meaning-giving — may, when combined, point with some assurance to that being or directed power which is man's peculiar endowment. This being is the power to bring the non-human into a human world, to convert chaos into culture, and to leave the traditions of world and culture to succeeding generations as places where they can begin their own efforts to render their human being manifest.

## §20. *Kinds of Truth About the Self*

Perhaps now the suspicion is growing that this section will interpret self-experience into the model X-subjectivity-Y by identifying X with the complex, variously functioning, and possibly unified self, and Y with the roles offered by a special social structure to which one may correspond and within which one may experience one's powers. It seems, however, unnecessary further to elaborate the relation of role to self. Emphasis instead will be placed upon use of the model by the self in coming concretely to know and to be itself. If X be the self, then Y may be some other actual or ideal person whom the self may come to resemble. X, thus, will bear a living truth-relation to Y. I want to point out here that lived truths of this sort are different in important respects from the more familiar sense of truth about the self and that they are developed in the course of the non-objective dialectic of which subjectivity is the condition.

We begin by returning to the analogy which pairs off self, subjectivity, and truth or correspondences with the source of light, illumination, and objects illuminated. In this analogy the source of illumination, the bearer of subjectivity, is identified as the self. The self as source of light is an expression of a truth about the self and designates a way in which one enters into one's own experience. But the one who so enters is not an object and is not illuminated as objects are, only by reflected light.

One is immediately aware of one's subjective powers as originating roles and as directed from a center. Subjective powers, one's possibilities, seem to radiate from this center like light. This spontaneous identification of the self as the one who is seeing — whatever may be seen — is one level of truth about the self. It presents the self as felt unreflectively. Some phenomenological observers have even localized this spontaneously experiencing and experienced center at a point which is normally just behind and between the eyes.[9] Convictions concerning its location are important here only in testifying that the experience is common and intense.

Just as one's body is felt and interpreted not only as the center of the environment but also as possibly or actually acting upon objects within the environment, so the self is experienced not only as the center and source of one's subjective powers but also as actively "outside" and busied about objects and also about the self as if it were an object in the world (cf. §6). That is, a person experiences himself reflectively as an object illuminated. He is the audience at his own drama as well as the hero. One sees oneself as engaged in premeditated actions, actions which are reflectively interpreted and judged. The self thus seen has certain properties in common with objects. In sum, then, the metaphor of light suggests that the self is twice involved in illumination. Like a source of light, it is luminescent; also, it receives illumination from light reflected back upon it from objects. The relations between the self experienced as source of subjectivity and the self as something made perceptible by that subjectivity are not always perspicuous. Let us unravel them to the extent of observing the ways (i.e., the two kinds of dialectic) in which these two kinds of truth about the self are initially formed.

Consider first the more familiar kind of seeing. One comes to know oneself objectively by use of the kind of sight with which one comes to know objects. A person feels guilt; he observes his behavior and notices that the guilty feeling prevents his reacting realistically in certain situations. So he seeks to understand this feeling and to control it. Or he examines his behavior empirically for the purpose of subjecting it to a science. In either instance a person examines himself, according to our metaphor, as something made perceptible by light rather as objects are made perceptible. These kinds of empirical and practical truth about the self are altogether familiar.

We shall not linger over them but turn to a kind of truth which is prior to them. Hence we ask: How does the self come to use its subjectivity in one way rather than in another?

Concrete truths about the self, like other truths, are discovered only by those who discipline themselves to the task of so using their subjectivity. Becoming a member of a culture entails receiving training of this kind. Such training proceeds first at a quite basic level of communication. It proceeds by a process of inner imitation, a formation of the person upon the model of that other, the mentor, who is imitated, or upon some ideal which the mentor offers. Thus, a possible other, whether he be visualized as some particular person or as a generalized other, is internalized by that concrete kind of communication which is called introjection. [10] In this manner, individual human beings come in some respects to correspond with each other. Here we speak of selves as being like each other not only as bearers of subjectivity, but also in that they tend to use their subjectivity to make similar decisions, to unify their experience in a similar manner, to respond to events in analogous ways. But they are not like each other as things are like or as things are like the images or concepts of them; rather, a correspondence holds between human beings in respect to their powers and dispositions for feeling, seeing, interpreting, and producing. To imitate a person and to assume his valuations is first to be a self, secondly to learn to use one's subjectivity spontaneously and reflectively to see in the way he does; this is to become sensitized to certain sorts of correspondences rather than to others. It is to acquire rationality of a distinctive sort. To see oneself as like other persons, thus as true to them or as corresponding with them, is not to see oneself as an object is seen but is rather to identify oneself with an habitual way of seeing and so to see the world as one's peers see it. This is the sense in which one's subjectivity is true of, or corresponds to, that of one's peers.

Before the reflective illumination of truth relations in the world, before the communication of such relations to some material, there must already have been a communication of some definite disposition, of some preference and direction of vision, to the one who sees. This initial specification of one's subjectivity directs, limits, and shapes common possibilities for seeing in a world, for grasping a context, for perceiving and manipulating objects. The source of human subjectivity is not a self which may be said simply to possess subjectivity; it is that which possesses it in a certain limited and partly controlled or controllable ways. To learn to direct one's power in one way is to resist directing it in another. Hence, a possessor of human subjectivity is a limited being, a self, which can see itself as like other selves and as more like to some of these than to others. Thus a person acquires an initial bent or is initiated into the style of a culture. Thereafter he decides and acts in a manner illustrative of that culture. He is a convert to it and accepts its assumptions, valuations, and mores—its rationality—without question.

This inner imitation of the other, this absorption of the valuations, the way of seeing, the prejudices of a culture, determine character structure. This determination occurs spontaneously and largely at an unconscious level. Psychologists make allusion to the deep and far reaching effect of this level of character communication when they remark that not what the parent says but what he *is* educates the child. Societies depend upon the mutual introjection of its members for reproducing the uniformity and stability upon which their identity depends. Thus, a society is able to lay down ahead of time even the ways in which a person may change.

A culture exhibits a common subjectivity. It offers its members certain rather definite and object-like ways of feeling and interpreting themselves. Once one has gained entrance into the average world of a culture, one may set oneself to experimenting with the roles which are meaningful within that culture. The discipline of acquiring a profession, of learning to play a role, can be visualized as a process of forming oneself upon a model. The student of physics, for example, is trying to learn to think like a physicist; he is shaping himself as well as producing a product. He is likely to recognize that by the exercise of his art he may both contribute to satisfying the needs of the society which fathered him, and may also use the roles presented to him as so many possible ways of coming to understand and to be himself.

In truth, then, one lives at least two correspondences: one with the vaguely imagined other of one's culture, through whom one becomes a spontaneously acquiescent member of one's own society; the other with the idealized role-player, the specialist on whom one more or less reflectively moulds oneself. Finally, and perhaps only rarely, in consequence of critical self-experience, one may transcend cultural patterns and, following the dialectic of human growth, make manifest a hitherto hidden possibility.

A culture, of course, is always already chosen for one. But it is reasonable that one's primary role in it should be chosen with some explicit care for the self. And, in fact, it can often be observed that, at least in youth, the time when one is most self-aware and reflective, the time before custom has staled the impetus to self-discovery, a principle determines the choice of role and the way in which, through trial and error, roles are played. I express this principle as the principle of unity. One will be most aware of those possibilities and possible roles which are in harmony with that which is past and acquired and which are also expressive of the anticipated self, the self drawing one on from the future. This anticipated whole self, indistinctly apprehended in one's own felt possibilities elicited, perhaps, by some recent course of events, may point to an interpretation available in the array of roles provided by the social structure. But this anticipation may also require a new and creative use of interpretation. In either instance, there is ideally a correspondence or unity between these felt possibilities and the interpretation, so that the self-interpretation expresses what there is to be expressed. The world-symbols of the West tend to draw the non-objective dialectic toward this wholeness, even though this unity or wholeness is often presented as something the more difficult to reach the closer one seems to approach it. In any event, this unity is not the sort used in ordinary counting or measurement. It is, rather, a quite special sense of unity, the unity of an individual person who can lay claim to uniqueness (cf. §33). Thus, we may suppose Gauguin's decision to drop the role of petty bourgeois banker and to take on the life of a painter brought his sense of himself into harmony with a new kind of role and thus released new powers and achieved, perhaps, a more inclusive unity.

It must also be added that this unity may only infrequently be realized or even sought. A person's position in life, the social structure, the ideals of a culture may forbid it by forcing him into a role wherein he cannot "follow his own argument" but must model himself on some alien figure. Such conditions are a source of the feelings of guilt and alienation of which so much has been made in our time. My points, however, are these: first, whatever the kind and degree of unity attained by the self, subjectivity is its necessary condition; second, this subjectivity acquires its unified, limited, and individual character, in the several respects in which it can vary, in consequence of the self's moulding itself upon those concrete and ideal others with whom it enters into relation.

The process by which this growth and unity are achieved has been termed the non-objective dialectic. This dialectic arises as the consequence of mutual influence exerted by feeling and reflective interpretation upon each other. Some stages in the non-objective dialectic of obligation were followed in the preceding chapter. Consider a contrasting example: Shakespeare's Macbeth became aware, in response to the witches' prophecies, of new possibilities within himself, feeling them as incipient royalty, a nascent omnipotent power. Yet he recognized King Duncan as the limit upon his power, still a limit — so his wife assured him — which he could remove. Acting upon this self-interpretation, he murdered Duncan and immediately experienced insupportable guilt. Eventually he interpreted this guilty self as little else than

> a tale,
> Told by an idiot, full of sound and fury,
> Signifying nothing.
> (V, v)

The readers of, or witnesses to, this drama are persuaded by the poet's skill to identify themselves with the protagonist and to move empathically through the action with him. Their incipient lust for power, their possible criminality, finds visible and concrete expression in the actor facing them, and they see these possibilities lead inevitably to a self-destructive end. That in the viewers which is akin to Macbeth is made manifest. They become, as it were, living metaphors of the Thane of Cawdor (cf. §33). And thus their experience of themselves and their sympathy with others is enlarged.

By means of the dramatic technique a playwright persuades us to love, struggle, fear, sense guilt, rejoice, and understand with the protagonists of his plays. The sympathetic viewer of the play could say to himself: I am related to my feelings, though without overtly acting, as Macbeth is to his actions and probable feelings. Mimetic communication is achieved by way of this kind of living analogy. The viewer's own horizon is widened by this temporary and experimental correspondence with others. A person thus sees what it is like to be an other; or better, he sees what he would be like were he that other. In this manner, the non-objective dialectic develops within subjectivity and takes for its aim the manifestation of the complete self. The non-objective dialectic is the means by which one's individual humanity is made explicit. The truth about the self is not merely a theoretical truth, it is the truth of a concrete correspondence of one's subjectivity with that of another person or with some idealized other. And this correspondence is what one is, at least for a time. If, then, one is obligated to pursue the truth about oneself, then one is also obligated to discipline one's subjectivity to the rationality which can successfully pursue this truth.

Although the self cannot be defined as any definite sort of thing, still we observe that at the least it has certain powers, powers which are limited by its world and which enable it to feel, actualize, and specify its indefinite possibilities, its subjectivity, by imitating another person and so to discipline itself to becoming an embodiment of his culture.

If this suggestion seems reasonable, that the self is formed by means of self-interpretation, and that self-interpretation proceeds in an important part by way of a mimetic communication with other persons, then the notion of person must be enlarged if justice is to be done to the range of self-interpretative behavior. For example, the Pueblo Indians, engaged in the famous Taos Deer Dance, must be conceived to be impersonating the gods or embodying cosmic forces.[11] "Life is a mystery play. Its players are cosmic principles wearing the mortal masks of mountain and man. We have only to lift the masks which cloak us to find at last the immortal gods who walk in our image across the stage."[12] In such a spirit Christian man strove to imitate Christ. He was enjoined by his world-symbols, not to a literal imitation, but to the achievement of an unimpeded and direct communication with the source of his being. And in a not totally dissimilar spirit, technological man strives to imitate the machine, that to which he looks for his salvation and his completion. His accuracy in acting and thinking, his impersonality and objectivity, his certainty of future conquests, are gifts of this deity. By such mimetic activity the self grows and acquires the rationality which is valued in his culture and which can deal with his environment. Occasionally, too, this rationality is bent back, through critical self-experience, upon itself, seeking thus to understand, perhaps to redirect, this process of the growth of the self.

A person's subjectivity, thus moulded, provides the means of access to roles, to nature, and to others. More especially, it provides access to the self. We turn next to utilization of this means of access to the self and to the kind of action and event which exhibits the self to the self in its movement to maturity.

CHAPTER V

# ACTION AND THE SELF

§21. *Access to the Self*

D espite failure to discover a definition of the self, we still have to deal with the need to determine the limits of human life and the desirable direction of development and to do so in a manner which will be useful for taking the measure of technological culture. I propose to search for these limits within the kind of change to which human being is subject.

It is difficult to see how any significant change is to be wrought in a person except through action. Projected human change which stops in fantasy, wish, or even conviction is unfulfilled and exists only in image or as prediction. Undoubtedly, Socrates of the *Republic* was correct when he observed that the journey out of the Cave toward the ideal good was properly completed only by a return to the concrete life of the market place whence it had begun. But if the limits of life are to be discovered in human action, surely they are not to be sought in the infinite detail of one's passing hours nor among the intertwisted incidents in the tapestry of social living. They are rather to be sought in the pattern of the action through which the self changes, perhaps changes radically. This development of self has already been described as the effect of the non-objective dialectic. Not the least common form of this dialectic is exemplified in the action through which a role, once accepted, comes to be altered.

We raise the question here concerning the pattern of action or the plot which reflection can discover in a man's effort to change himself and to render his felt possibilities explicit and concrete. It is not difficult to show that a rather simple and general pattern is discoverable in efforts toward self-interpretation. I want to exhibit this pattern and to show that when conceived on a large scale it also reveals the limitations within which any man may pursue self-understanding and make manifest his possibilities. It may be suggested that knowledge of such limits is common: the Ten Commandments offer an illustration. The Ten Commandments, however, were delivered to Moses on Mount Sinai: they belong to a particular culture. Since we desire to assay technological culture, we have need of a view of human limitations which transcends such cultural particularities. We must, therefore, exert ourselves to discover these limitations in the human condition itself.

The question first concerns the kind and structure of the action which is effective in producing that change which renders one's humanity explicit. A few suggestions have been made in the preceding chapters. It has been suggested that this action is a reflective interpretation of one's felt possibilites, that it is of the nature of trial and error in respect to role playing, that it is dramatic, that it is a (non-objective) dialectic, that it is animated by the need and obligation to make the truth about one's humanity explicit, and that an essential part of the communication of truth takes place mimetically. This chapter is intended to exploit these pointers. Let us first take note of the general direction in which this inquiry will lead.

The ever present obligation to pursue the truth about oneself points like a Janus-face in two directions. Obligations have their roots in the past. They develop out of the indefinite and incomplete character of human nature. Man's incompleteness arises from the fact that his initial awareness of himself—his self-presentation—is, like the presented element in perception, relatively indeterminate. And the understanding of this self-presentation is not fixed and laid down in advance but must grow through some process of both spontaneous and reflective self-explication. In other words, a man is incomplete in that what he is is not already settled and defined but depends in great part upon his felt possibilities, his self-interpreting decisions, and the manner of life which follows therefrom. The truth which he pursues lies in the future not only in that he is always in pursuit of it but in that the truth about himself which he does acquire refers him to future contingencies for its realization. The truth about himself which he has reached testifies to nothing so universally and so clearly as to the fact that man is obligated to act in terms of a truth which he is incapable of attaining completely. He is subject to a fate which he fails clearly to grasp. He feels himself limited but does not know just where his limits lie. This native obscurity, this on-tological fault, is exhibited in his every act. Thus, his obligation to search out the truth about himself places man in a position of strain (cf. §15). As respects the present essay this strain is most acutely felt as a difficulty which it places in the way of discovering a standard for measuring human value, for instance, the value of a scientific and technological culture. If man has no fixed essence, if he cannot in principle determine *what* he is, how can he evaluate or take the measure of his contemporary mode of life or of the cultural ideals which he is asked to pursue?

Under the sway of the strain in which he lives, man experiments with different self interpretations and ways of life, diversifies his culture, and breaks with even the most firmly fixed tradition. Is the result merely a series of equivalued cultures and traditions? Does man change merely for the sake of change? I think he need not. The history of his changes demonstrates certain regularities. One of these regularities is repeated in the lifetime of perhaps every person as he moves to maturity and seeks ways of using himself in the culture and world at his disposal. This pattern of change is easily observed. It forms a constant theme of literary and other works of art. Perhaps it is most clearly exhibited in the drama, especially the

Greek tragic drama, and may most easily be drawn from this source. Some mention has already been made of the drama. We shall, though, need to sketch the moments of the pattern of this action more exactly and to indicate their practical function and their relation to the great world-symbols which have guided life and thought in the West. There is reason then to believe that drama offers the means by which a man may undertake a critical interpretation of himself and discover what powers, what identity, what limitations may be his. But of course it need not be so used. A man may prefer to remain in ignorance of his powers or of his identity; or he may contrive to substitute some flattering fantasy for the insight which otherwise he might have achieved. Nevertheless, whether fantastic or realistic, the roles which a person chooses as his own constitute steps in his self-interpretation.

We shall direct ourselves in this chapter toward examining a pervasive structure of role-playing, one related in its earlier history to a background of ceremony and myth and by this route to common experience. Then we shall endeavor to be led by this clue to some insight into the self, the kinds of changes to which it is subject, and its ensuing fate. We hope to draw therefrom some indicators which will point the way to self-understanding and to the limits within which this understanding can be pursued.

## §22. *The Structure of Role-Change*

A role has a career in time. A man accepts a role, identifies himself at least partially with it, seeks to execute it, and then eventually terminates it. We require, then, terms for the description of the power of role playing which will do justice to the power of choosing or originating roles, the capacity of sustaining and developing them, and the power for appropriately terminating them. And we desire to see how these movements are executed when they are executed rationally.

With this intention we might turn to seek among the experts in the manual or professional arts and their careers for some means to penetrate to the structure of role practice. I fear, however, that this turn would be unfruitful since the exercise of these arts has become so complex in modern times, so fixed in specialized institutions, so involved in changing patterns of the division of labor and of labor shared with machines, so penetrated by scientific or semi-scientific views, beliefs, and prejudices as well as by various external conditions of practice, e.g., economic conditions, that their structure is obscured. This structure, however, can be expected to become explicit in the theater, for there the self and its roles and role-change are taken as themes. I turn, therefore, to the theater as to a laboratory where the matters in which we are interested are most likely to be exhibited. A preparatory discussion of this topic has already been offered (§ 15).

Dramas are various and complex. But if one were to limit consideration to the ancient Greek tragic dramas and to the basic *conditio sine qua non* of their structure, one would be likely to agree with Aristotle of the *Poetics* that a drama represents a complete action, and a complete action is one having a beginning, a middle, and an end. Aristotle further points out that the beginning of a drama is an episode which is not logically preceded by another; its meaning requires no preceding event, but it does point to a succeeding event. A middle points in both temporal directions, for it is an episode logically requiring both temporal directions, for it is an episode logically requiring both a preceding and a following event. An end logically requires no sequel; its meaning becomes evident within itself. The end event achieves this finality by returning in some manner upon the initiating event so as to form a kind of circle, the figure regarded by Aristotle and by other Greek thinkers, as the complete or perfect figure.

That which marks an intitial episode as a beginning, requiring no other to precede it, was reasonably seen to incorporate a decision on the part of the protagonist.[1] This was a decision to undertake a certain role within the given context. Typically the requirements for discharging such a role are laid down by the structure of the society within which the protagonist lives and as a rule are communicated mimetically (cf. §20. At the beginning of Aristotle's favorite play, Oedipus resolves to act the part of the just king of Thebes and wise father of his family. The middle represents his effort to carry out the obligations imposed by this decision or self-identification, by actually playing the chosen part in an acceptable manner. In general, this portion of the play consists in the struggle — a heroic and failing struggle — to retain the chosen identity in the face of mounting evidence against it. Finally, the end moved to an insight into the justice or injustice of the chosen role. It achieves completeness by reflecting back upon the initiating decision, so bringing the action full circle; it displays the manner in which this initiating hypothesis concerning the hero's identity is measured by the facts and by the hero's fate as it unrolls in actuality.[2] For Oedipus, the evidence revealed that he was not truly the King of Thebes, nor was he appropriately the father of his family. The roles he had chosen for himself could not be maintained; his existing self was not mirrored in his more complete self.

His eyes had been so fixed upon the outer environment, in the endeavor to obtain power over it and to retain his position in it, that he had failed to turn them inward in inquiry into his own being. In the end, though, he was forced to turn his sight upon himself and so to change himself. Note that such action, leading sometimes to radical human change, is quite different from action as often understood. Often human activity, like that of a falling stone, is supposed to unfold in homogeneous, objective time. It is said to exhibit one incident after another in a linear series. Even some of those rebellious inheritors of classical humanism, the Existentialists, tend to conceive of human action with reference to *engagement,* a cause, which leads through a succession of events toward final success or failure. The drama which we have chosen as a guide, however, exhibits human action developing in dramatic time and revealing a cyclical movement in which the end moment returns upon the beginning as the protagonist grows toward the whole self. It thus exhibits a characteristic movement, perhaps *the* characteristic movement, of common experience and of the non-objective dialectic.

The insight of a drama, which gathers the beginning up into the end by means of the middle episodes, seldom reaches an affirmative grasp of the self. Consider that *Oedipus Rex* has long been recognized as the tragedy of the young man who resists the transition into adulthood and is forced by fate into recognition of this failure. Similarly, Shakespeare's *King Lear* is the tragedy of an old man who decides against making the transition into old age but is dragged by fate to accept the consequences of his time of life. As in these plays, so typically in the tragic theater, the insight is negative and expressed recognition that the hero had in fact misidentified himself and his role at the outset. Ironically, the knowledge his suffering won was the knowledge of who he was not. Such an end cannot be appreciated without some understanding of two related elements: fate (anagke or Moira) and hybris.

Some notice of hybris and of the ontological fault have already been taken (§ 15); here we turn to fate. In fate we recognize the obscure, even irrational, horizon against which the Greeks visualized human life. It was a mysterious power or else a demand of the Gods to which man necessarily had to conform. Some men fate leads; others she has to drag. But whichever, there is no escape from fate. However, men generally, the tragic hero in particular, being possessed of hybris or arrogant confidence in their own power, are blind to their own limitations. The tragic hero believed himself to know or else to be able to determine his fate. Inevitably events revealed to him his blindness, but only after the tragic sequence of events does he grasp his failure to turn his eyes inward, to follow the non-objective dialectic, and to pursue self-understanding. Thus, tragedy is an ironic struggle against fate, a struggle which reveals the blindness, the incompleteness, in which it was initiated.

My emphasis upon the negative character of the dramatic insight should not be taken to exclude a few apparent exceptions, e.g., Socrates, whose positive insight into his questing

function in Athens and in relation to himself seems continually
to have been reconfirmed in the Platonic dialogues. A martyr
might be defined as a protagonist whose decision concerning his
identity and role is reconfirmed repeatedly, until no further
confirmation or disconfirmation is possible. This reconfirmation can be interpreted as testifying to the protagonist's
harmony with his fate; however, the protagonist's confidence
that this reconfirmation by fate will continue is dependent upon
a firmly held conviction which usually differs from that of the
majority. Socrates expressed his faith as touching this point in
various myths (e.g., *Phaedo* 107-115A). But it must also be
remembered that Socrates held this faith with Socratic
ignorance.

By writing dramas in cycles of three, the Greeks recorded
their confidence that the protagonist could move himself closer
to harmony with his fate (e.g., the *Orestia* of Aeschylus).
Aristotle regarded the insight which terminated the dramatic
movement as cathartic; the cathartic insight purified the
protagonist of elements extrinsic to his essence and brought him
closer to harmony with his rational and ideal self. In the West,
however, under the influence of Christianity, this confidence
disappeared. Man's hybris was regarded as having affected his
very nature, so that no individual, however clear his human
insight, however excellent his skill and courage, could bring
himself into accord with his fate or save himself. Thus, no finite
number of merely human dramas could be regarded as com-
pleting man's deficiency. Dramatic insights alone cannot be
supposed continuously to purify the protagonist and to bring
him thus into harmony with a definite and clearly envisaged
human essence. Nevertheless, each man is fated to be saved.
Though cathartic insights may prepare a man for salvation, still
each has to depend upon some power beyond himself to
complement his ontological defect. The question concerning
the identity of this power has provided the perennial problem of
the Western world; I have noted that Western cultures have
offered a series of beliefs about the nature of this power (pp.
61f.). These beliefs range from faith in a Supreme Being to

confidence in technology. Although many hold this latter confidence with something of Socratic ignorance; many also hold it naively, there being no evident and commonly understood standard by which to measure its truth.

At the moment we do not need further to compare worlds or to examine literary variations on the theme of decision, struggle, and insight as exhibited in tragedy, comedy, and tragi-comedy.[3] We need only observe that the adoption and enactment of roles in our world, perhaps in any, approximate to this pattern. Such is a characteristic movement of the non-objective dialectic. Thus, a young man feels it within his power to become a physicist and decides upon this career. He then undertakes to perform in this role by submitting himself to the various disciplines and obligations demanded by graduate study, laboratory practice, and professional life, only at length to discover that this way of life is "not for him." Perhaps he finds that he lacks talent or interest, or he may conclude that the person he is becoming is too widely at variance with what he feels himself more truly to be. It may seem to him that such a self no longer counts for much in the scheme of things. Perhaps he cannot find a job as a physicist, or maybe he merely becomes too old or too ill to play the role any longer. If now, he had genuinely identified himself with this professional role, then the insight that the part is no longer for him approaches tragic insight and requires of him the sacrifice of the role-self which he had initially chosen. After this insight into his misidentification of himself, he may decide for another profession and enter again in on the tragic cycle. By means of this pattern of action, the power and impotence inherent in the self are made manifest.

It may even be suggested that the tragic cycle provides a kind of measure (in the broad sense of the term) of the self. "Measure" is etymologically related to "month" and to "moon." The reference is to the very early division of time into months by means of the waxing and waning of that sky god, the moon. The dramatic measure of the self, though, refers not to the division of its time into convenient periods but rather to the appearing and exhibiting of its powers and then to their withdrawal in impotence until another cycle is initiated.

In sum, than, a role is a culturally defined, or at the least a socially permitted, schema of self-interpretation. For present (non-objective) purposes, it is understood to be a connected pattern of action through which the self enters into and becomes effective in a general or specialized context and finally terminates this activity, whether from reasons intrinsic to that activity or imposed from without. To play a role, then, is to seek to develop and to conserve a correspondence between certain felt possibilities of the self and a reflective interpretation of the self expressed in the terms of a social function or perhaps in more significant and fateful terms. Often there remains in the background of this pattern of action the vague and mythic presence of world powers which once gave a place and a meaning to human life and in whose honor dramas once were staged. If any one structure can be held to be drawn from common experience, I should suppose this structure to be the pattern of role trial and change of role. Human living is a succession of roles adopted to express the inevitably different life periods, the several visions of self, the several functions in which one is involved, the differing needs and aspirations of one's culture, or the various demands of the times. In any event, the self playing a role is interpreting or making explicit his grasp upon his own possibilities. He is manifesting his being in action having the structure of decision, struggle, and insight.

## §23. *On the Identity and Non-Identity of the Self*

As primitively felt by men, environing nature and especially human nature are sensed as profoundly threatening and fearful. Still it seems that both natures, the outer and the inner, can safely be regarded and even controlled if, like the ancient Gorgon-headed monster, they be viewed indirectly as if in a mirror. The mirror which later man had devised is forged of symbols. In our time we may flatter ourselves that we have utilized this mirror with considerable success as respects the environment (cf. §§18, 30). We have learned rather effectively how to map nature's routines and changes in symbolic systems which manifest the being of nature and bring her under our control. The same, however, cannot be said of our success in reflecting upon human being. Like Oedipus, we tend to ignore the inner and to direct our gaze upon the outer nature. Hence we have by no means improved the system of symbolic rituals and sacraments which mirror human life and through which we might gain self-understanding and self-control. Indeed, many thoughtful observers believe that we have in recent centuries lost ground in this respect. Our greatly improved skill in practicing the objective dialectic is unfortunately matched only by our loss of sensitivity to and skill in the conduct of the non-objective dialectic. Thus, finding in the ancient pattern of tragic action a key to self-discovery is, I suggest, scarcely an anachronism.

The names "subjectivity" and 'rationality' have been given to man's readiness to enter in on this pattern of change. Acknowledgment of this power is wordlessly expressed by accepting the decision that one's own way lies in being such and such a person, i.e., in adopting a determinate role. Such a decision leads through struggle to a culmination in the final insight that this initial decision was defective, perhaps altogether erroneous, and hence demands either adjustment or radical change of a sort which affects the self and may even alter the culture, perhaps it eventually is manifested in the world itself.

An important difference between comedy and tragedy is to be understood with respect to these two possibilities. Tragedy culminates in a radical alteration of identity, an alteration that affects the structure of one's subjectivity, whereas comedy demands only readjustment of role within the same culture. Now my point is that a recognition of either kind requires the prior conviction that one possesses a valid if inarticulate self-awareness, or feeling which is at the least a recognition *that* one exists. This apprehension or felt possibility is initially thought to be adequately expressed or interpreted in the decision. The final insight is like a comparison of this decision with the self so far as the self is revealed by the test of action. The conclusion drawn from the comparison between the choice of role and the self revealed in its action was often expressed in ancient times as an admission of failure to achieve a correspondence between one's fate and one's decision regarding oneself. (One failed to recognize the true self.) In modern times we note that a person is often urged to a choice of role by the structure of society and by expectations nurtured in family and local life. Such conventional choices are frequently remote from the feelings, desires, needs, or talents of the individual; hence, it is scarcely surprising that an unhappy consciousness of self is not an uncommon outcome of the dialectic initiated by a conventional choice of role. In any event, the recognition of this unhappy outcome is also recognition that an inarticulate awareness of self is used as the measure of such a role, of its appropriateness or the reverse (cf. §8).

This mode of access to the self, as if by way of a feeling, an intuition, or an inarticulate yet powerful conviction, not always subject to adequate expression, strongly suggests that the self is felt to be more than a succession or collection of roles. The self is not only the possibility of other roles than those actually selected, but were not, and roles that may later be chosen, but the history of self always offers more than a hint that any role whatsoever will finally prove to be inappropriate (cf. §26).[4]

The lesson to be learned from this pattern of the non-objective dialectic with respect to role-playing is simple. It follows from the usual negative character of the final insight that the self is not wholly to be identified with any specific role. Such was the insight of *Hamlet,* of *Oedipus at Colonus,* of *King Lear.* The fixity and object-like character of roles as they are offered for one's choosing by social institutions no doubt should predispose one to anticipate this negative conclusion. A person accepts this conclusion, however, with understandable desitation, for if the self is never any role, what can it be? Where will its limits lie? Chapter II used the comparison of the self to poetic language which always said more athan a literal translation could compass. Can this poetic comparison be rendered more intelligibly? But first we must make our way through an objection.

## §24. *Toward the Meaning of Self*

Perhaps the difficulty encountered in the attempt to give literal expression to the meaning of self is indicative of just the sort of thing which language analyists believe to be the symptom of a linguistic or conceptual muddle. Professor G. Ryle offers, by way of an illustration of this kind of supposed muddle and the cure for it, an account of someone showing his university to a visitor.[5] As they walk around the grounds, he remarks, "Here is the library building, there is the laboratory, yonder is a lecture hall; this is the refectory," and so forth. After the tour is finished, the visitor asks, "Now that you have shown me the library, the laboratory, and so forth, I would like to see the university." Here Ryle makes his point. There is no "university" over and above its several components. To suppose that there is, is to be misled by language into a mistaken use of categories; it is to suppose that because some nouns name particular objects that all must do so; hence, "university" must name an object, although in this instance a queer and invisible one, a pseudo object (a ghost in the university). Rather than posit such an illusory class of objects, one should recognize this hasty generalization concerning the usage of nouns. Thus, happily, the illusion that there are such queer, invisible, and universal objects can be cured by a more careful examination of language.

Could the conviction that there is a self in addition to the roles played in daily life also be a muddle, subject to the same analysis and cure? But let us examine Ryle's example with a little more care. Let us ask to be shown around the library. The same sequence develops. "There are the stacks; here is the catalogue; yonder is the circulation desk," and so forth, we shall be told. Then we may also ask, "But where is the library itself?" By a most reasonable analogy to the preceding argument, the question is to be regarded as naive. The linguistic stupidity of the questioner has again led him (us, this time) into a silly "Platonism." So, then, we have committed the category mistake again in our use of "library." The word names no object over and above the components of it. What of "book"? By the same argument a book, too, may be dissolved into a collection of components which has no identity apart from these components. Any object word could thus be shown to end up with no referent of its own. By successive applications of this method all names turn out to be names only of illusions or of pseudo objects. We are not surprised when another philosopher of this nominalistic tradition, playing Hume's game rather more seriously than Hume, identifies the self as that which learns to use the first personal pronoun correctly. Here the problem of nominalism emerges; nominalism leaves nothing to name, unless it itself is false. As respects our inquiry, this kind of linguistic analysis of "self" produces more of a muddle than the one it was intended to cure. A more useful guide is offered by an analogy drawn from biology.

It is well known that most of the cells of an organism are specialized to perform particular functions, but every organism normally retains some cells in non-specialized condition. These are the cells which may be used to clon, or to reproduce the original organism a-sexually. They contain the whole organism in possibility. Now I am suggesting that roles are to the self rather as specialized cells of an organism are to the un-specialized ones.[6] But in order to use this analogy it must be recalled that roles are not things like cells in an organism; they are acts, or rather they are ordered patterns of acts which when executed, produce a product or otherwise effect some complex function in the body social. These are specialized acts. Are there, by way of contrast, some non-specialized acts, or patterns of acts, in which all men are engaged and which prepare the way for, yet stand in contrast to, the specialized ones? If so, then perhaps these non-specialized acts, like the poetic word, will point more directly to the self which originates and plays roles without being or being reducible to roles.

I believe there are such acts, and I propose that they be

determined by recourse to the criterion of replaceability used earlier (pp. 77 f.); if a person can be replaced in an operation or event, then he is operating as a role-player; if not, then he may be functioning as a self. And thus, likewise, we distinguish between the self's possibilities which enable it to play some specified role and those possibilities without which it could not be a self at all.

Now all the roles which ordinarily come to mind and which one plays throughout his life, can, no doubt also be played by some other. For instance, professional, leisure, and game functions can always be performed by another. This point is aptly expressed by a remark heard from time to time in Naval circles to the effect that no officer or man, however efficient and important, is indispensable. But further, more intimate roles such as those of father, friend, councilor not infrequently are fulfilled by another. Finally, as we have already noted, the most intimate of roles, the role of spouse, while often averred to be unique, does not always prove to be so. Thus, in respect of roles a number of individuals are or can be operational equivalents of each other. Indisputably, however, there are certain actions which cannot possibly be performed for one person by another. Let us observe what actions these are and whether they can in some respect lay claim to being indissolubly related to self and self-determination.

If every act, involvement, or role in which one engages and which could be performed for one person by another were laid aside as being roles or parts of roles, hence partly independent of and perhaps unessential to the self, the philosopher would rather quickly find himself concerned only with such events as birth, life, death. For being born, living, and dying are acts which cannot possibly be performed by one person for another; hence, they are not roles. Also they are acts in which everyone must be engaged; they belong to common experience and are essential to it. They are the acts in which both the being of the self and the limits of this being must originate. Hence, they can be expected to reveal the self. Perhaps, too, they will indicate the limits within which the non-objective dialectic should move.

I shall now make trial to see whether they do reveal something of the self and of these limits.

## §25.  *The Essential Human Events*

Surely birth, life, and death are essential to any human existence. But *what* are these events? What interpretation is to be assigned them? I want to show first that these events contain and can exhibit the self very much as the non-specialized cells of an organism contain and might reproduce the whole. Secondly, and in the next section, I want to indicate that the account of the self reached by way of an interpretation of these events will exhibit the limitations within which the pursuit of the truth about the self, whether theoretical or concrete, is to be pursued.

It may at first seem that these matters have already been discussed, for birth, life, and death appear to be merely another version of Aristotle's notion of a complete action, only on a larger scale. For birth is surely a beginning, and death just as surely an ending. And life time is the mean between the two, related forward to one and back to the other. On the other hand, birth is not a beginning in the sense in which a decision is the beginning of a human action. Likewise, death is not an ending in just the way in which insight terminates an action. Birth and death are clearly, then, not parts of the pattern of role-playing. Birth at least is more like a concrete condition for a life of role-playing. In sum, the similarities and differences between decision, struggle, and insight on one hand, and birth, life and death on the other are evidently complex. Let us consider them more closely.

It may seem strange to pick birth and death out as essential human phenomena, since all animals and even vegetables undergo them. In identifying human being so closely with such events we may seem to let slip just that which renders one's own experience typically human. Yet so soon as we attempt to capture that which was let slip, we recognize how difficult it is to divine what could be meant by a human experience of birth or of death. The first is forgotten if ever it were known, and the other is not yet. Yet if these phenomena present that which is to be interpreted if one is to know oneself, then doubtless all the experiences of life would point to them in some manner. Our problem is to discern the sense of their reference to a human self and to the limits of his being.

Let us begin with birth. Just what is human birth? Clearly we are all born in the usual manner. Birth is commonly said to be a coming into being by way of an emergence from another being, the mother, as a certain specific sort of individual having its own relatively independent life. This age-old conviction is linked with the belief that man is an animal, although a thinking animal; our method, though, requires that we lay aside such conventional convictions. Hence, we must ask: can this biological process be typically *human* birth? I think not. In the first place, as just observed, this process is common to all animals. Moreover, no newly emerged animal can quite properly be said to be a language using animal, even if by convention it belongs to the human species. But more significantly, human birth is an emergence into a human family, tradition, language, set of attitudes and a history. In short, human birth is emergence into a human world and into human time. A difficulty lies in understanding the meaning of "emergence." For ordinarily and literally one emerges *from* one place *into* another. Emergence, that is, is a transitional movement. However, birth, understood as a first "emergence" into the human world, cannot be emergence in the literal sense. We can scarcely think of biological birth as a movement from a non-human world and time into a human world and time. The time-honored images of birth, e.g., passing through a gateway, are not literally descriptive of human birth. And one can scarcely seriously maintain that the experience of birth is integrated into the whole of the experiences which are identified as one's own. Doubtless the same difficulty is to be seen in the psychiatric interpretation of birth as a trauma. A trauma is a wound which happens to a being who is already there to be wounded and who can be aware that he is being wounded; but the infant a-borning, who is not exactly already there and who has not learned to distinguish itself from other beings, cannot be supposed to be able to suffer a wound in just this sense. We cannot exactly say who suffers the birth trauma.

Thus, the biological process of physical separation from the mother cannot without paradox be identified as human birth. Human birth in a more proper and intelligible sense is the beginning (at whatever chronological age) of the recognition and voluntary use of human powers. Birth is the beginning of a possibility of interpreting or pursuing the truth about the self within human limitations. Birth in the biological sense can be human birth only in a derivative and figurative sense.

Our purposes require the identification of some experiences which will point to the significance—the reflective interpretation—of the phenomenon of birth. I suggest that a repetition of human birth is undergone by anyone acquainted with the solving of problems of all kinds, from mathematical ones to those involving human relations. A few illustrations will make the point. Consider a crucial one: the "Silent Concerto" by John Cage. In this concerto the pianist comes out onto the stage. He elaborately arranges himself in front of the piano, prepares his hands to play, and remains motionlessly so for three minutes. The notes which the audience confidently expect do not issue forth. The "concerto" is silent. What is the meaning of this production? Sometimes it is taken as a joke and passed off as *avant garde*. And so it may be in one sense. However, there is another and more philosophical interpretation of this concerto. Its silence may be taken as communicating quite seriously and musically to the audience. I believe it communicates the audience's own possibility for enjoying the music. The concerto presents the audience precisely with its own preparation for the experience of a concerto. So, of course, does any pianist who seats himself at his instrument and seems to prepare himself, actually waiting a few moments while he and his listeners "get into the appropriate mood." With the usual concerto, though, the music quickly follows this preparation; and the musical awareness and expectation of the audience is forgotten in the music itself. With the "Silent Concerto," however, this experience of possible musicality, as it might be called, is the whole communication. This felt possibility, this readiness for music, is something which only a person can achieve. It is that without which there can be no experience of music. A tone-deaf person cannot achieve it in fact; a dog cannot achieve it in principle. I would like to refer to this apprehension of one's power for hearing a concerto as a kind of awareness of the possible emergence of music. What John Cage has opened up to his audience with his "Silent Concerto" is an occasion for their witnessing the dawn of their own power to hear a concerto. This power is subjectivity, that which converts a collection and succession of sounds into a musical whole and is suggestive of the agency which converts a collection and succession of roles into the partial unity of the existing self, the self actually in movement toward the vaguely felt whole self, its fate. In consequence of this subjectivity a biological organism is a human being. Husserl spoke of this subjectivity as the wonder of all wonders; and I have noted its analogy to the poetic and dramatic dimension of language, the power of giving meaning to the non-meaningful.

There are many other phenomena of just this kind, experiences which point to a similar event of birth or of the beginning of this characteristic human power. The blank canvas, or the canvas presenting nothing more than a single color, framed and hung, seem also to invite us to an awareness of our felt possibility for experiencing painting. Such a "painting" has in fact been exposed; I refer to Kasimir Malevich's "White on White" (now owned by the New York Museum of Modern Art). Likewise, an empty theater, or the silently revisited scene of one's childhood loves and fears, offer other illustrations of this kind of readiness or awareness of possibility.

Many people look back upon their childhood as islands of self-awareness which become connected and definitely related to one's self when one emerges into adulthood, as dreams are merged into one's self-identity upon waking. Likewise, some persons experience and have recorded their sense of initiation into the power to be human, the beginning of selfhood. Illustrations of becoming aware of specific possibilities or roles are, as it were, little births; they point to and help give meaning to this birth of the whole self, which is the beginning of the individual's power to be human, the preparation of the man for self-interpretation and commitment. This sometimes dramatic recognition of the capacity to be an individual self—one's entrance on the stage of life, so to speak—has often been experienced and portrayed in literature. I cite a single instance which comes from an autobiographical work by C. G. Jung. He writes: "Suddenly for a single moment I had the overwhelming impression of having just emerged from a dense cloud. I knew all at once: now I am *myself!* It was as if a wall of mist were at my back, and behind the wall there was not yet an 'I.' But at this moment *I came upon myself*—Now I knew: I am myself now, now I exist."[7]

The same experience would be had by anyone who underwent a radical or even a partial reorientation in life, for instance, a reevaluation of the world as in a religious conversion, or in the event of a more or less complete transformation of one's being or mode of being. Learning to be at home in a once strange culture is an illustration. Man is the being which can be born and reborn in such a manner. He is the being which can sense ahead of time capacities for a kind of being—a *vita nuova*—and can determine himself to pursue or not to pursue his interpretation of this possibility. Thus, it seems to me to be evident that the understanding of birth as typically human birth is gathered not from the biological event of separation from the mother, but rather it is to be gathered from such experiences of rebirth as the first felt awakening of a power of some kind, the awareness of a possibility or of a "power-to," which renders it reasonable to suppose that one could enter into and play a specific role.

Birth in this sense often occurs in a time of crisis, a time when one discovers that old patterns of living are inadequate. This is a time when old habits of interpretation are recognized to be insufficient for resolving one's problems. Then one is like a little child again, facing the unknown, dependent upon others. Such a time, therefore, calls upon one to discover some new and unused possibility hitherto dormant within the self. This new possibility must be made manifest, matured, and set to work to reorder one's life or to resolve the problem. This kind of a birth which is the discovery of a new possibility, of a "power-to," helps to throw light upon the primary sense of human birth.

This light, however, is no more than analogical. Human birth is also unlike this dawning sense of the power to be human, for the latter always includes awareness of the fact of decision. Within limits, one can choose the role one will play. Birth, however, is beyond one's choice. Heidegger has described it as being thrown. In being born, one is thrown into a culture, a time, a life. What family? What race, sex, tradition? What native language will be mine? These questions are already settled by birth. They call not for decision but for acceptance of this dependency. Participatory democracy plays no part in determining the answers to them.

In sum, I find the awakened sense of possibility, of power to achieve something, to be the original analogon which leads us to recognize something like a human birth in the biological event: not *vice versa*. What I am suggesting, in other words, is that the customary view of birth has confused the literal with the figurative. Human birth in the literal sense is to be identified, as our illustrations have indicated, in the sense of awakened power to experience, to choose an end, to solve a problem, to play a role. It is the emergence or transition from an old into a new self. Birth in this most proper and literal sense has subsequently been extended to include the quite paradoxical instance of biological birth, that birth which in the first instance means life. Birth in this latter and figurative sense is not a fully human and freely chosen event as respects the one undergoing birth; it is not literally an emergence at all but an altogether mysterious "movement" from nothing into subjectivity and time, a movement which we are able to grasp, if at all, only through metaphor. It is the beginning of the possibility of being oneself; it is not exactly like deciding upon a new role. Taking this figurative and biological description of birth too literally is responsible for the trivialization of biological birth and for overlooking an important source of insight into the human self and its nature. Specifically, it overlooks man's dependence, the fact that the basically important decisions—what tradition, what talents, what culture, what myths, what language will be mine?—are already made for one and woven into the basic design of one's fate.

I want, now, to point to one of the consequences of the failure to understand birth and the dependence which it connotes in its initial and literal context. The present interpretation of the peculiar human quality of birth was approached, it will be recalled, by way of an interpretation of John Cage's "Silent Concerto" and Malevich's painting, "White on White." It was said that these works could awaken one to a recognition of one's musical or painterly capacities. There is, however, another interpretation of these works which points in a quite different direction and elicits another and perhaps dangerous aspect of birth. Birth is, after all, an awakening of power, a beginning; but a beginning of what? Such a power may bode either good or ill in its beginning, and thus it may be a threat as well as a promise.

When Kandinsky viewed Malevich's "White on White" he is reported to have remarked that the painting had failed, for it was still canvas and was still white. His point is clear enough. He understood the painting to be an effort to transcend the conventions which bind painters and seem to prevent their exploitation of aesthetic possibilities. Similarly Cage's "Silent Concerto" may refer us to the apparent limitlessness of the possibilities for the musical filling up of silence. And in fact, in the initial indicating of capacity or possibility, limitations upon possibility are often not in evidence. We are still, in our period of history, sufficiently close to the Romantic interpretation of the world and to its Christian background to be aware of the drive to transcend all limitations, to flout conventions, and to break old moulds, in the expectation of transcending the human and acquiring a magic power of salvation. Fascination with possibility leads easily to this *strebende nach dem Unendlichen*. The Romantic person, in love with infinity, constantly seeks to loosen the ties imposed by the world: the Romantic artist is traditionally impatient of the restraints placed upon him by dependence upon the materials and technical requirements of his art. He is intoxicated with a sense of creative inspiration, of the power to originate. This fascination with the possibilities opened up by birth induces the Romantic person to attempt to become free of all limitations and to ignore human dependence; it has led the sphere of art, some believe, to chaos, the political world to the brink of barbarism, and the economic and industrial worlds into an increasing commitment to a supposed omnipotent technology. Birth, which is a discovery of possibility, does not at the same time immediately reveal the appropriate limitations upon this possibility. Although birth is an unmistakable exhibition of human dependence, it is never more than an indefinite promise of fulfillment. Hence, an investigation into the intrinsic limitations placed upon the possibilities of fulfillment needs to be engaged.

Now I think it obvious that intrinsic limits are not to be sought in commonsense and convention. Conventions can change, and when they are not changed but are conserved beyond their usefulness, they become incitements to just the Romantic rebellion to which I have alluded. Neither can we assume that the sciences discover limits, for their function is to operate within limits already set for them. Furthermore, when the sciences are applied technologically, it is fairly clear that they tend to violate human limits even more surely and regularly than does Romantic art. They aspire to extend without bound human control over the environment and over other human beings—it is ironic that a seemingly sober scientific technology has come to represent a Romantic impulse carried to its uttermost extreme. Is there not, however, somewhere an intrinsic limitation upon this impulse to technological power? No doubt we should return back again to the phenomena themselves to discover there some clue to the nature and location of the boundaries to the possibility which our ever-widening horizon of human birth continually opens up to us.

Among the three phenomena with which a man's fate is linked and which are surely basic to the human being, birth, life, and death, the last is most evidently related to human finiteness. For death is a limitation upon life which is in effect organically entailed by birth itself. In analyzing and describing the typically human sense of death it may be that we shall come upon some aspect of it which can function to define the limits of human possibility envisaged in birth.

We begin by observing, as we did when examining the human sense of birth, that one does not experience physical or biological death. If an "experience" is a *human* event, and a human event is at the minimum an event which one can include within his time, i.e., can both anticipate and recollect, then biological birth and biological death will have to be denied the status of being experiences. For one can neither look forward to one's own biological birth nor can one look back in recollection upon one's own biological death. We, therefore, resort again to analogies drawn from likely experiences to get at the sense in which death is the final determinant of human being.

A difficulty is presented by the fact that death in the biological sense, unlike birth, occurs under so many forms. There is the martyr's death, the functional and triumphant conclusion of a purposeful life. At the other extreme, our traffic and crime reports testify to the frequency of accidental death, the senseless ending. Accounts of some suicides read like a senseless ending to a life become senseless. Moreover, some deaths occur following lengthy intervals of coma, long after all hope of the patients ever becoming himself again have had to be relinquished. Is the biological end of such a comatose organism death in the specifically human sense? Surely not. We might ask whether such a comatose person ever experiences the death of the body. But more properly we must ask whether any meaning at all can be given to the phrase "experience of death." And again I would reply that it cannot be an experience, since it has no retrospective dimension. The illustration of the comatose individual renders the lack of this dimension obvious.

Thus, it is not even clear how a specifically human death is to be identified. Let us, then, for the moment accept tradition as our guide and attempt to disentangle the figurative from the literal usages of the term "death."

The kind of death which has always been most prominently indicated as characteristically human is that which seems to be most nearly under the control of, or to fall within the intention and design of, the person who dies. I cite again the classic instance of such a death, Socrates'; upon his deathbed he gathered up the loose ends of his life into an intelligible whole which effectively brought his biography to completion by carrying out the intent and purpose of his whole life. Such a death is martyrdom. It is a witnessing to, a manifesting of, the self which one is. The phenomena which present death in the characteristically human sense are the phenomena of this kind of fruition. But clearly a fruition of this kind is not experienced solely upon a death-bed. Death in this sense can be with us at any time and may illustrate what Plato had in mind when he spoke of death as the end of the practice of philosophy.

Such an ending or death, I suggest, is best exemplified upon occasions of insight; it is the intrinsic end or the intended completion of a period of preparation. The single perception of the pattern of a whole concerto, for example, would be an instance of musical insight which terminates the concerto. When, it might be asked, ought the "Silent Concerto" to be ended? No doubt just when the audience gets the point. Also in the dramatic realm the achievement by the hero of a drama of an insight into himself and the meaning of his role offers a clear instance of such an ending. It realizes the possibilities anticipated by the initial decision and thus rounds the action out to its close. It brings to an end a period in the life of the protagonist. So too, in the management of more specialized rational problems, there may come a moment of total grasp of the data when every item falls into an intelligible form. Henri Poincaré offers an excellent mathematical illustration. Upon stepping off a bus in Coutance, the solution to a problem concerning Fuchsian functions flashed suddenly into his mind. That task was ended; he had only to note his solution and to write out the details for others to read.

Death understood thus, as the end or completion of an epoch of living, is always occurring. It is reasonable always to expect it. Doubtless Plato was correct; a good ethical injunction is that we should live in pursuit of it. The moment of death is one of the moments defining life and the rhythm of life. Acceptance of death in this sense is recognition that the insight into human being which gives it meaning and direction is the intrinsic limit or end of his being. One possessed of such a belief reasonably regards the self as the being obligated to pursue insight into himself. Death in this intelligible and dramatic sense is the recognition of the fulfillment of possibility. Any moment of life can become such a moment of terminal insight. Again, I want to insist that the intrinsic end of a human action is an insight into the meaning or logos of that action, and this end is death in the literal human sense. Death understood as an essential element in human being is the power of reaching such a terminal insight.

Biological death, however, is something unlike as well as like such a terminal insight. We may approach this unlikeness by recalling that martyrdom is exceedingly rare, and that the usual terminal insight of a tragic drama is negative. The protagonist learns through suffering the tragic action who he is not. Perhaps this negative note offers a clue to the sense of biological death. This death is not merely an end in the sense of a negative insight; rather, it is the negative of insight, the limit of the possibility of any kind of sight. Frequently in the tragic drama the hero meets his biological death. Hippolytus of Euripides' play by that name, is killed in the end. His end, though, is not meaningless; it illuminates the one-sided character of the decision which initiated the drama. It shows that he cannot play the part merely of the lover of Artemis, as he had supposed he could. He, of course, does not give expression to this meaning; however, his parents, as it were, express it for him and for the audience. He himself is beyond this negative insight in the nothingness of biological death. He has passed into

> The undiscovered country from whose bourne
> No traveler returns,

and has been absorbed in its non-being.

My point is that mere destruction of the biological organism by extrinsic causes is not a human death at all, unless this destruction be given a meaning by being interpreted on analogy to the specifically human sort of ending or completion to which our illustrations have pointed. And when it is shown that the average man is a being who can conceive of death while denying its existence, the implication is reasonable that such a man has not succeeded in giving a meaning to his biological end. He has not succeeded in grasping his own life in its completeness and in its limits. He has not found the courage to maintain his recognition of the limits on his power to be human.

With these reflections on the senses of birth and death in mind, we return briefly to consider life again. At least, having recognized in birth the beginning of human possibilities, and in death their end, we recognize in the phenomena of life that which links the first of these events to the last. Specifically, human living must be the movement of developing self-awareness from birth, through its history, and toward its death. Perhaps we might describe life as a maturing grasp of the significance of birth and death. Thus, it must also be marked by a maturing of the power to feel and to interpret such significances Life might, then, be said to be that which disciplines subjectivity to rationality. It is the struggle to reach and retain the rationality which sees these significances. Life is, therefore, the movement between the awakening of human possibilities and their completion in their intrinsic end. It is the present moment of strain between beginning and end, decision and insight, obligation and truth. It becomes concrete in the roles which one adopts as one's own and plays until they are seen to be no longer viable. Life is a series of discoveries that one's past interpretations of oneself are false. No one is ever just what he thinks he is or has resolved to be. This conclusion is the point to Sartre's example of the waiter in a pub who, contrary to his efforts, cannot be what he is, that is, just a waiter. One's possibilities overflow any such role and falsify any attempted interpretation of self exclusively in terms of role or project.

The typically negative conclusion to role trial argues that roles are no more than accidents of the human being, personas, masks, which are worn for a time, then when they are played out, are laid aside. All decisions about the self lead through catastrophe to other decisions. Perhaps this generalization is true of the several definitions of man with which Western history has made us familiar. Each describes the typical role-player as he views himself at some epoch or other of our history. But the self itself is, rather, the power to accept, to play, and then to end such a role; it is that which enters into being with birth, lives out a life of role trials, and ends by laying life aside when all such identities are lost in death.

Human life, then, is lived in dramatic time, the time whose common rhythm is marked by the cyclical return with insight upon one's chosen and presumed identity. This is a temporality which urges one to self-confrontation and invites one to engage in the non-objective dialectic. But likewise, as will be seen (§37), it is a time whose articulations can be neutralized by certain cultural or world views, and whose differentiating moments can be evened out or routinized by interpretations which devaluate the human.

Is there any direct experience of this self, the germinal cell of one's humanity, the unity of all the different roles which one may ever play, together with the events of beginning and ending? Just as the "Silent Concerto" is contrived to provide an awareness of a possibility, of one's readiness for music, so there may be certain experiences which bring to the surface of consciousness one's total intention to be human, one's readiness or power to infuse the endurance of time with the rhythms of concrete human living. I cannot say what these experiences are; they seem sometime to be mysteriously chosen. Efforts to give them public content seem regularly to fail, unless perhaps in the works of poets. Some aver that a discipline of contemplative silence will bring them to the fore; I do not know. My observations indicate that the self is most usually exhibited through the "power of the negative." The self is primitively presented in the sense of readiness or "power-to." We become clearly aware of a possibility or "capacity-to" upon the occasions when it is frustrated, when in consequence of some forced or blindly chosen role we come to see that the basic and non-objective dialectic of our existence has led to unfulfillment. Doubtless St. Augustine gave appropriate expression to this negative and problematic character of human being in the phrase chosen as a motto for this book: *"Quaestio mihi factus sum."* We return, thus, by another route to a conclusion already suggested (§16): that the life of the self is basically a career in self-interrogation, at least it is such a career until death, the final question, is reached. And this event marks irrefutably, whether acknowledged or not, the finitude and incompleteness of the possibility to be human and to give meaning. What we have fixed upon here is not a definition in answer to the question about the self. But the interrogative form has been taken as the form of man's existence. As the human being becomes actual in the large rhythm of birth-life-death, so its subordinate phases exhibit the same movement in the repeated interrogative structure of decision, struggle, and insight. Precisely this interrogative structure gives meaning to our guiding declaration: that a man's primary obligation is to inquire into the truth about himself and to embody this inquiry in action.

My turn in this section has been to relate the self essentially to the great primary events of birth, life, and death, the acts which cannot be performed by proxy. The self is readiness for birth, life, and death, and man's fate is to be subject to these primary events. But the meaning of birth, life, death, in the usual acceptance of these terms has been judged to be figurative rather than literal. We cannot really say that they are unified in a single clear insight into the whole self. Rather we hold open the question about the nature and unity of the self. Nor can we speak of these events in literal terms. This figurativeness deserves to be emphasized. In the more familiar illustrations of beginning, middle, and end which occur within the course of life time, for instance, in the illustrations adduced above, the middle is given its meaning by the events of beginning and ending. Whereas, on the contrary, when the whole of a lifetime is taken as the middle between biological birth and biological death, then this middle becomes a source of analogies from which we seek to take meanings for the initial and biological birth and for the final death. The pattern of role trial and change of role, which characterizes one's life history, offers a source of analogies — but only of analogies — for the interpretation of birth and death. This reversal of the source of meaning should be sufficient to maintain awareness of our ignorance and consequently of the appropriateness of an interrogative attitude toward the basic phenomena of man.

In sum, these phenomena, birth, life, death exhibit the primary possibilities of man, his elementary being, and point to his intrinsic limitations. This being itself is the power to be born, to live, and to undergo death. These possibilities are instantiated recognizably in the adoption and rejection of roles. But the role is not the being, though it uses this power. Birth, life, and death mark the greater temporal rhythm of man; initiating, executing, and terminating roles are the moments linking the beginning and ending of this greater time. The man desirous of fulfilling the primary obligation to know the truth about himself will, one would suppose, seek to utilize the minor periods of his life, those exemplifying the pattern of decision, struggle, and insight, as a source of analogies which will help to illuminate the essential and common events of human being.

## §26. *The Limits of Self*

It will be appropriate now to attempt to reap what we have sown and to summarize the preceding comparison and contrast between, on the one hand, the beginning, middle, and end of an act of role-playing and, on the other, birth, life, and death of a person. For it should by now be evident that these two are closely related, though they are by no means identical. Such a summary may yield at least a guide to determining the logos of the self. This guide will express the limits within which the attempt to determine the truth about the self can appropriately be pursued. Again it should be noted, although every culture develops limitations which bear upon and regulate both public and private life, our undertaking to evaluate technological culture requires appeal to limitations which transcend particular cultures and approach something like world conditions and have their basis in the concrete conditions of human life generally rather than in some particular culture. Such limitations were all but discriminated in the preceding section and now need only to be made explicit.

Biological birth, we concluded, is not a beginning of the self in a dramatic sense, rather it is the beginning of a subjectivity, of the possibility of human self-hood. But this beginning of the individual is indefinite. It embodies no initiating decision, rather it plunges the newly born into a pre-selected family, tradition, culture, and language, upon which he must depend and within which he will make his decisions, acquire his character, and dispose of himself. As the infant organism, a possible self, emerged from the mother, so the self is continually emerging from association with others, from the matrix of culture, and within its world. Indefinite in himself, the individual is set upon the way to becoming his definite self by mimetic communication with others. This dependence upon others and upon one's culture determines the initial bent of one's life. The essential event of birth, then, is the beginning of dependence and entails, we must remember, the continued dependence of self upon factors which are not oneself.

Biological death is the end of one's possibility to be human. It marks in all respects and constantly the unsurpassable limit of one's time and of one's power to be oneself. It completes one's span of time but does not complete the self. The ever present possibility of this end in death entails the finiteness of selfhood. As the self cannot but come somewhere upon the end of its possibilities, so each of its decisions, each project, must enclose within itself its intrinisc limits. The self is, so far as our knowledge extends, a movement between nothing and nothing, between the not-yet humanly possible and the no longer humanly possible. The self is an intrinsically limited set of possibilities.

We observed, too, that life time, the period for maturing and aging, is employed as the time for playing a variety and succession of roles. And a role, we recall, is a related and socially involved set of actions, having the structure of decision, struggle, and insight. The characteristic outcome of all these roles is the negative insight that the self is adequately mirrored in none of them. Ultimately the self is non-identical with any of the type-identities offered by a culture, whatever that culture. Such is the nature of the self's patterned movement between nothing and nothing.

We may, thence, draw the conclusion that the self is dependent, finite, and only negatively known. This description of the self is mainly negative and hardly constitutes a definition; nevertheless, it is surely an indispensable guide to discovering such a definition. It lays down the limitations within which the reflective interpretation of the self is to be pursued. To exceed these limits in any way, however well concealed, whether in feeling, expectation, practice, or theory, is to invite eventual self-destruction.

The practical use of this guide is illustrated by referring again to role-playing. For consider that if a person is a physician, then he must constantly recall just what the role of physician requires. If he forgets these requirements, then he may play the role in an unjust, distorted, or destructive manner; then surely he will run afoul of the law or of his patients and be forced to suffer the consequences. This identifying recollection of one's role is the condition for playing it, whether this role be in a kinship system, a social structure, a work relation, or any similar context. Likewise, a person who is endeavoring to be himself must constantly recall the limitations to being a self, even if this knowledge be only negative, the knowledge of what the self is not. He must, then, constantly recall his dependence, finiteness, and the negative outcome of role-playing (his non-identity). With this guide in mind and within these limits, he may pursue his completeness by way of the non-objective dialectic with some modest hopes of success.

The emphasis, no doubt, must be placed upon modesty, for the limits indicated here are severe. It is not surprising that men, while contemplating such limitations, were easily convinced that a super-individual power would have to come to their aid were they to maintain any confidence of success in attaining completeness of being. However, great culture heroes of the West have displayed confidence of no mean order in facing recognition of their dependence, finiteness, and non-identity. Consider Alfred the Great, St. Joan of Arc, Giordano Bruno. The example of such heroes indicates that one can always in some fashion become aware of the primary obligation, accept it, and seek to discharge it. When a person does so, I shall say that he has the courage, being guided by recognition of his limits, to make his humanity explicit. For short, I shall merely say that he is courageous.

It is quite easy to fail to be courageous. The ways of guilt are many. One need only avoid reflection upon oneself and accept without question some conventional mode of life, or one may identify oneself so completely with some function or role that one fails to become aware of oneself as something other than the role-player, or one may disguise one's future and imagine oneself to continue somehow infinitely in a non-terminating role, and the like. In effect all of these ways of avoiding oneself are ways of imagining the self to be fate, or even superior to fate; they are hybritic, and invite the revenge of fate. They are ways of refusing to recognize the dependence, finiteness, and non-identity with any role which are inextricably linked to human life and action in any culture, even in any world. These ways of avoiding oneself may require and exhibit considerable rationality, but none develop that kind of reason which enables a man to discharge his primary obligation.

## §27. *World-Symbols and Fate*

The limits just indicated, bounding human action and self-knowledge, are negative. The question must be raised whether there is a positive guide to the making of courageous decisions in the pursuit of insight into human being. The view to be proposed here will merely render explicit a procedure long familiar to common experience but often hardly noticed. This is the view that guidance toward completeness in the manifestation of one's being is effected by world-symbols.

World has been described as the structure of relations and values which unite a man to himself, to others, and to nature; it is indicated by the traits common to a group of cultures. World is expressed implicitly in the daily life of man and society; one's awareness of world comes to explicit expression in certain dominating symbols of creation, destiny, and destruction. The extent of these symbols is suggested by the large and inclusive emblems which refer to a certain institution in one's culture. The British royal family, the American national flag, are such cultural symbols. They unite, presumably, the patriotic feelings, national memories, political aspirations, and ideals of a nation. Of world extent, however, are the great symbols of creation, fall, salvation, and ending: consider Yeggdrasil and the Ragnarok, Ouranos and Gaea, Dionysus, the Christian doctrines of creation, Christ, and the New Jerusalem, Galileo's mathematicizing God, the modern infinite cosmos inviting spacemen to make and man craft for transporting their species to unpolluted planets. Such symbols shape and direct cultures. The more effective ones engage men's responses on many levels.

Important world-symbols for the Greeks were the myths and rites associated with fertility gods and goddesses, e.g., with Demeter, Dionysus, Orpheus. By means of these rites a man is related to nature, its seasonal changes and its productivity, to himself as a participant in this yearly renewal which may secure survival for himself, and to others who equally assist in the rite of festival and reap its benefits. The post-Classic Western world preserved many elements of these rites, but its most influential achievement in the development of world-symbols is probably the Christian doctrine of God and of man as the image of God. I recount here only the main outlines of this world-symbol and note only one of its more recent transformations.

God seen by Christianity appears as an ideal community of complete persons engaged in the creative manifestation of infinite power. I refer, of course, to the doctrine of the Trinity of persons (actors) who use the one divine substance (power) in three functionally different ways: the first person is the creator of nature; the second person is the Wisdom or intelligibility in which the cosmos was created; and the third person is love which moves creation to embody this Wisdom,[8] and gives expression to the directional factor of the being of one's world. Man is the defective or fallen image of this Trinity; he is moved, however, in friendly concert with others to seek to restore this image within himself to its complete nature by way of creative activity. Man's fall, however, has rendered his nature so indefinite and so remote from its divine origin that, unaided, he would be at a loss how to understand his present plight or how to direct himself toward regaining his completeness. To supply this need, the second person of the Trinity became incarnated as a human being. The Incarnation thus became the central world-symbol for Mediaeval culture, the symbol through which a man's existing (and fallen) self could be related to his complete self (his destiny and salvation).

Mediaeval theologians and philosophers interpreted Scriptural accounts of Christ as exemplifying a doctrine of one person (or actor) uniting in himself two natures or substances. One substance (power) is the divine and immortal being; the other is human and mortal. This duality of the divine and human is also represented in the two trees which, according to the author of *Genesis* (II, 9 ), the Creator God caused to be placed in the Garden of Eden: the tree of life and the tree of knowledge. Their unity in the one, the "light of the world," has been symbolized by the sun whose rays bring light, life, and growth to the world, but whose being is cyclical and inclines also to winter darkness and death. Adam and Eve, given being by

God yet desiring more power, ate of the tree of knowledge, seeking thus to be as a god, but in consequence of this self-misidentification became instead fallen humans, afflicted with an ontological defect, and aware of impending death. Christ in his human nature is associated with this death, with crucifixion and the tomb, but he unifies this mortal being with the divine. The primary symbol of Christianity combines both the divine tree of life (the cross was called "The Tree" in Mediaeval times) and human death and sacrifice (the crucified man). Thus, the Crucifix is the symbol of Christ combining in his one person both the divine and the human natures. The whole self is these two become one. Christ as world-symbol is, thus, the image of the complete man.

The Christian teaching, however, does not suggest that these two natures can be united in any man as they are united in Christ. Rather, Christian doctrine insists that the human and divine are incommensurable, for man's nature owns to the ontological defect. This doctrine adds, nevertheless, that upon the initiative of the divine and with the cooperation of the divine and the human, a man may eventually attain completeness.[9]

This Christian world-symbol points to man's relation to nature as to a gift, to his relation to others through love and through the common endeavor toward completeness, and to himself by holding up the vision of unity with his forgotten or hidden diviner self.[10] This doctrine was embodied not only in architecture, icon, and music but also in rites in which all could participate. These rites which mirrored man's complex and paradoxical nature operated with powerful mimetic effect to shape the selves of the participants.

I have pointed out that the Western world has preserved in its several cultures different versions of three important Mediaeval convictions: that the self in its present (fallen) state is not defined, perhaps not definable; that the self is indefinitely remote from its most desirable and complete state; and that approach to this completeness requires some extra-individual aid. The Christian world symbol — the Christ image — dealt for a time effectively with the condition described by these three convictions. There is a question, though, whether the present Western world possesses anything as effective as this Mediaeval symbol. That the religious version of this symbol is no longer current, is more than obvious. Is it, nevertheless, still effective in some other cultural variant? If not, what sort of relation is presently maintained among self, others, and nature as the self moves through changes and attempts to deal with its fate?

A person always finds himself already in a world and in a culture not of his own choosing. There are no premises from which this placing is an inference. This placing, nevertheless, forms that life and gives it the primary convictions and interpretative principles which determine its view of itself, of its birth and death, as well as of others, and of its world. Thus, it determines in a general way the kind of life which the self will seek to lead. It offers the self a limited set of roles to play. While playing them, while disciplining his subjectivity to the rationality required by his culture, the person is still the victim of natural events and disasters as well as of his times, of their political and social vicissitudes, of their wars and peace, of their anxieties and values, of their visions and rejections. Then, finally, accidents, illness, and death, so often untimely, seem to be a meaningless invasion and disruption of whatever order and meaning he has been able to discover or to establish. This irrationality, the unaccountableness of these crucial events, is that to which I give the name of fate. Man's rationality is inclosed within this larger irrationality; his life is a little clearing within the darkness of fate.

To recognize that fate is not merely an accident nor a series of accidents to which one must passively accommodate oneself, but rather that it is an irrationality which pervades the very power to be human, is surely an important step in the pursuit of the truth about oneself. At the same time this fate is human fate; the power to be human touches it also and seeks to give it meaning. Interpreting fateful events, giving them meaning, finally referring them to world-symbols, is the human way of dealing with their irrationality.

Perhaps no other Western culture is in the long run as optimistic as Mediaeval Christianity which sees the darkness of man's fate as in reality the blinding light of his divine destiny. But all cultures attempt in some way to render this darkness less intense by interpreting it through world-symbols. Even technological culture has, in the course of its recent development, thrown up several such symbols, the most familiar being the image of someone in a machine, the slim blond in a speedy red roadster, the hard-hat in a Gantry, the cosmonaut in a space module. These images seek to assure us that we, too, may complement our feebleness with a machine.

In all of these images, though, man and machine are still distinct and relatively independent of each other. A far more advanced symbol strangely recalls, rather parodies, the Christian symbol of the man-god, the two substances in one person. I refer to what has been called a Cyborg, an intimate unity of man and machine which recent developments render imaginable. Indeed this strange combination seems to demonstrate the basic identity of man and machine long suspected by those faithful modern souls who rejected the uneasy truce of Cartesian dualism. Of course man and machine have often been intimately conjoined. Eye glasses, prosthetic limbs, and "pacers" to control heart beat are illustrations of machines closely related to men; the Cyborg, however, is supposedly connected to more powerful machines of more varied capabilities and more directly (e.g., through more or less direct connections to brain centers).[11] Direct electronic communication between brain and machine unify the two, man and mechanism, "in one person" in a manner strangely reminiscent of the mediaeval world-symbol. The resultant "man" is presumably, in possession of indefinitely more power and can control other people and his natural environment to an immeasurably greater degree than man without these saving additions. Possible acceptance of the Cyborg, or of something like it, even if only on a popular level, as a symbol guiding men's actions and forming their ideals should be sufficient to provoke critical consideration of such symbols. A hopeful circumstance might be for philosophers as well as poets to concern themselves about the discovery and expression of world-symbols appropriate to their time.[12] For world-symbols are the speech of the world to man and indicate to him who he his, where he is going, and why. This essay, though, goes no further than to suggest that such a symbol might well reject the Cyborg and instead incorporate the image of the existing self aware of its poetic dimension, meditating the meaning of birth, life, and death and so discovering the limits within which it can redeem its alienation from the complete self.

Although the world-symbolical power of the Christian man-god has waned, the Cyborg, that prosthetic god, has not yet taken its place. Indeed, today there is no dominating world-symbol giving positive direction and urgency to the impulse toward self-completeness. Under such conditions and in this twilight of the gods, perhaps the search after truth about oneself can at least be negatively guided by recognition of one's dependence, finiteness, and non-identity.

# CHAPTER VI

# TECHNOLOGICAL RATIONALITY

## §28. *On the Role of the Scientist-Technician*

We shall in this chapter turn our attention upon a very important phase of the objective dialectic. This dialectic provides the means by which a man makes manifest and communicable to himself and to others the nature of the being usually called (physical) objects. As I am using the term, it includes the sum total of the means for reaching a knowledge of nature. The process is correctly called dialectical, for the relations between that knowledge and the means of obtaining it is a reciprocal relation. That this relation is reciprocal is easy to see. In an early stage of the history of this knowledge, the available knowledge rendered possible the construction of crude instruments (e.g., instruments of measure). Then these latter facilitate gathering more reliable data and the construction of more precise concepts; thereupon, improved instruments of measure could be developed, which in turn enable more exact data to be gathered, and so on apparently without end.[1]

We cannot in this chapter be concerned with the whole of the process of the objective dialectic but will confine our attention to that part of it which has achieved crucial significance today, *viz.*, the art and rationale of measurement. My purpose will be to demonstrate that a fundamental and essential activity of all science and technology, *viz.*, measuring, is a process of interpretation which successfully pursues certain kinds of truths but which, by reason of the languages involved, is incapable even of approximating to certain other kinds of truths, e.g., truths about the concrete self; although, a quite unusual usage of the language of measurement can be made to point to truths about the self.

I suppose it obvious that the average man of today is sure that the function of science is to study nature for the purpose of prediction and control. Many scientists agree with the average man on this point. If they are correct, then, in fact science serves industry, which again is to say that the scientist has become the scientist-technician. Let us for the moment accept this wide-spread view and proceed to ask: what is the role of the scientist-technician, or for short, of the scientist? Our question, however, will have to be more limited: what is the role of the measurer? And even here, we shall have to restrict our interests to those aspects of measurement which will enable us to determine some characteristic limitations upon this technique.

This restriction to matters relative to measurement is reasonable. It is evident that techniques of measurement have to be elaborated and applied in the data-collecting phase preliminary to the formulation of an hypothesis, in the phase of verification of the hypothesis, and at every step of the way in the phase of application. The techniques of measurement which yield access to the objects of the sciences also determine the meanings which objects possess for a scientifically disposed culture. Thus, an appropriate shortcut to an insight into a culture given over to technology is to be got by way of an examination of the kinds of measurement essential to technology. Of course, a study of innumerable other aspects of technology would yield important understandings, but the present purpose will be most expeditiously served if we confine attention to this shortcut.

A first point to note is that the role of measurer cannot be played by just anyone. The player must be disciplined to his part. The discipline necessary for this role presents a contrast to that specified in section 5 as necessary for much of the present philosophical undertaking. There, as I pointed out, a kind of partial epoche, suspending the results of the sciences and of scientific modes of interpretation, was requisite in order to approach common experience of the self and its world. Here, in order to understand or to practice the role of measurer, this partial epoche must be exercised in reverse. This reverse epoche can be achieved in two steps. (1) A general and standard attitude of impersonality or objectivity must become the dominating way of responding to objects. Since interest here lies in the invariants in the behavior of natural objects, the *way* the investigator himself feels about these objects (outside of the standard attitude of impersonality and objectivity) is irrelevant. Now since the non-objective dialectic begins precisely with feelings such as those now requiring to be suspended, we may say that just this non-objective dialectic must be placed to one side by anyone who enters in on the practice of science or technology. The scientific investigator always stands aside in his own person and becomes (ideally) the impartial judge of the data and of their relation to his hypothesis. (2) A most important part of the investigator's training must be directed toward acquiring certain standard habits and procedures in the use of his instruments. For example, in taking meter readings the investigator must learn to look at the pointer from the correct position, so that he gets the same reading that any other trained observer would get. This art of laboratory technique is extensive and more difficult to acquire than strangers to the laboratory usually suppose. Almost every scientific instrument demands arduous training and practice in its use. The investigator, then, is one who has become an impersonal and standard observer and who is skilled in the use of certain instruments. Only he, and not just anyone, is equipped to practice the objective dialectic or to undertake even the accessory role of measurer. (I am by no means suggesting that the practice of both the objective and non-objective dialectics is in every sense incompatible; but at the present under ordinary circumstances both are not practiced at the same time.)

In one sense the art of measurement is extremely old and very much a part of common experience. This art is simply an application of subjectivity or the power of seeing and interpreting as this was described in chapter IV (§§18, 19). Measurement, however, has now become a highly specialized form of reflective interpretation. One might think of the role of measurer in Platonic terms as a kind of Eros. Eros, it will be recalled, was an intermediary between the gods and men (*Sym.* 202Df), carrying messages back and forth between them. Now the mathematical Eros, similarly, mediates between the timeless heaven of intelligible form and the realm of perceptible objects (images) which are to be measured by reference to the timeless forms. Today, however, measurement is apt to be described as an application of conventionally defined units of measure and familiar operations for applying them. And analogously in the sciences, only the sciences usually refer to more exactly defined units of measure, to more complex and sophisticated methods of applying them, and to more carefully refined data than are encountered in ordinary experience.

Once the character of exact measurement is displayed, it should not be difficult to grasp the limitations intrinsic to the languages of measurement and thence to apprehend the traits of those aspects of nature to which measurement yields access. The role of scientist-technician, whose function is to deal authoritatively with that nature, will then be easier to evaluate. More generally, these considerations should provide some insight into the subjectivity which is disciplined by practice of the objective dialectic to technological rationality. We shall, therefore, be concerned with the general nature of measurement, with the specific sort of measurement essential to modern science and technology and also with its pathology. Finally, some attention will be given to contrasting the art of measurement thus described with a quite different "measurement" relevant to the man seeking insight into his own possibilities by way of the non-objective dialectic. The understanding of measurement, both of what it is and of what is is not, which the present chapter seeks to reach will provide a useful preliminary to the measuring of technological culture next to be engaged.

## §29. *On the General Nature and Use of Measurement*

The practice of the art of measurement so essential in modern Western culture could not thrive just anywhere. The modern vision of space and time, for instance, envisages both as eminently measurable. Modern space and time are spontaneously seen as quantities, infinitely extended, and divisible. In consequence of their continuity, connectedness, isotropism, and homogeneity, they are always measurable.[2] Objects or matter, too, are extended and endure; they are measurable by these spatio-temporal measures and by others which quantify matter. As Renaissance or early modern man understood it, the boundlessness of space and time, and the regularities of matter are a sort of reflection, upon a simpler level to be sure, of the infinity and necessity of God. Mediaeval man, rendered miserable by the conflict within him and by uncertainties regarding his own nature and destiny, sought to transcend the immeasurable distance from his wholeness by means of a reunion with the infinite creator God. But in Renaissance times this God acquired mathematical properties. The infinity of space and time became a symbol—a world-symbol—of the divine infinity. The lawfulness and accuracy of the movements of celestial (and other) bodies became a symbol of his perfection. Understanding these symbols was held to guide men to reunion with his source and thus to complete his being.

These world-symbols, however—infinite space and time, the celestial machine—are not related to man as the religious symbols were. True: Renaissance man is bound to this world for disciplining and for preparing himself for salvation; but the discipline and the salvation are no longer Mediaeval. This man is related to this nature by the relation of its mathematical intelligibility to his mathematical intelligence.

It is as if the world were specifically made in order to be intelligible to man. Men of Renaissance culture felt that in laying bare the mathematically intelligible aspects of nature they were thinking God's thoughts after him. Galileo believed he was translating into human mathematics the mathematical thinking of God written in the book of the universe. Kepler believed he perceived the signs of the Holy Trinity in the simplicity and mathematical character of the planetary laws. Montpertuis was guided to the law of least action by the expectation of finding in physical nature some reflection of the infinite perfection of God. Also Renaissance man was bound to himself and to others by the same bond of intelligibility. He conceived his function to be the discovery and development of sciences, for thus he expected to render nature intelligible. Others are his helpers in this, the great project of modern Western culture. Achievement of this project is to be the salvation of man. Indeed, man was regarded as rational in essence, a thinking thing, and the power of thought, *intellectus,* was held to be best exhibited in the doing of mathematics (cf. §16).

Modern cultures continue along the Renaissance path but turn aside from its theology. Thus, for modern man as for his Renaissance predecessor the infinity of space, time, matter, and even of man himself, are to be transcended by being rendered mathematically rational and intelligible. This same rationality eventually becomes the instrument for using and controlling nature. The means for rendering this intelligibility explicit (for translating it into human mathematics) is measure. What, then, is the nature of measure?

Music, poetry, the dance have since antiquity been mentioned conjointly with measure. Music, for instance, is divided into measures. That is, a rhythm of so many counts per unit measure is explicitly imposed upon the otherwise patternless duration of musical sounds. The consequence is that the duration of the sounds is limited in predetermined ways. Holding to this rhythm, it becomes possible for many persons to play or to sing the same music together and in harmony. The patternless duration of the sounds is brought by rhythm into the area of the common and the rational. Thus, that which merely endures is brought by this discipline into human time. In general, the operation of measurement is a process of bringing the variously and vaguely grasped into more clear and common possession. Thus, man and his world are reduced to rational order. Let us turn now to a more limited consideration of the order which measurement introduces or discovers in nature.

For our purposes a useful beginning statement about measurement will be one which will point out its similarities to other more familiar operations. Such a statement is this: measurement is an operation by which new knowledge is obtained from old. Immediately, though, measurement must be distinguished from the strictly mathematical method of obtaining new knowledge from old. For the mathematical or deductive method, the old knowledge is expressed in definitions, postulates, and rules of deduction. Application of these rules in conjunction with the postulates and definitions elicits new statements or new knowledge. Both the new and the old statements may be quite independent of experience. Measurement, however, is a method of obtaining new *experiential* knowledge from old. The method proceeds by the discovery and utilization of correspondences between what is well and clearly known and what is less well known. The old knowledge is relatively accurately grasped. Then if this old knowledge can be seen or shown to mirror or to correspond to something less well known, thereby the less well known becomes better known. Measuring, then, may be described as showing the correspondence of a class of well known or standard objects (the *mensurans*) to those less well known objects which are to be measured (the *mensurandum*). Thus, measurement utilizes a type of the truth relation as this relation was described in §18.

This general process is operative in many sorts of apparently simple instances of perception. Let us say I perceive this object in front of me to be a tree. The old knowledge is my grasp of what a tree is. I expressed this point earlier (§4B) by imagining myself to carry around with me a stock of standard representations one of which is the tree image or concept. The less well known item in this learning situation is the presentation now before me. I am not in total ignorance concerning this presentation. At least it occurs in a context with which I already have some familiarity; perhaps I am in a forest or in a situation where a tree would be no anomaly. This context functions as a guide or heuristic rule for selecting the standard representation which is likely or appropriate. I thereupon select the tree concept; that is I "see" that the presentation matches well enough with my tree concept. Consequently, I interpret the presentation as a tree; otherwise expressed, I intend it as a tree. The passive part of this operation is the reception of that which is presented. The active part is the selection of the tree concept or intention together with the act of matching by which I (by now automatically) recognize the tree in the presentation. The old knowledge is my prior grasp of what the tree is; the new knowledge is recognition that that which is presented is to be interpreted as a tree. Measurement, thus, is a kind of reflective interpretation; it establishes explicit correspondences between knowledge already possessed and that which is less well apprehended (in the illustration, a presentation).

An adaptation of this employment of subjectivity was developed in Chapter III where obligation was discussed. There the knowledge already possessed concerned the self as it should be — whether determined by feeling, by culture, or by a criticized ideal — and the self as in fact it was apprehended. The failure to observe a close correspondence between the two was interpreted as some form of guilt. Generally speaking, qualitative measure is performed in just this fashion; the definition or description of the object in question does duty also as a norm. Judgment of the degree in which a given object illustrates or fulfills the definition provides a qualitative measure of the given object (cf. §3).

The loosest form in which the seeing of a correspondence between elements or classes is exemplified is in analogies, whether used in ordinary or in poetic communication. I choose for illustration a metaphor contained within a quotation from *Macbeth*. Lady Macbeth had just urged her husband to wash the blood of the murdered king from his hands and so to conceal his guilt. Macbeth replies,

> No, This my hand will rather
> The multitudinous seas incarnadine,
> Making the green one red. (*Macbeth*, II, ii)

Macbeth is recalling that the blood of the murdered Duncan would redden the water in which he washed. But his mind leaps beyond the literal image of the bloody water; he feels that the hand which murdered the king discolors the life—giving sea, or rather the guilt in the soul of the murderer stains the waters of life, indeed the world itself. The analogy might be made explicit thus; as the bloody hand is to the reddened water, so the guilty soul is to "Neptune's ocean." Thus, by use of the form of correspondence, A : B :: X : Y, Macbeth measures or gives evaluative meaning to his bloody hand. The qualitative measure of the significance of the bloody hand is exactly this: overwhelming guilt. A Positivist critic might protest that the relation in the two instances—the ": "—is not similar; or expressed otherwise, the redness of the blood in the basin of water and the redness of guilt in this "sea" are equivocal uses of "redness"; hence, the correspondence does not hold. To maintain that it does hold is merely to exemplify once again the intolerable fuzziness of the poetic imagination and its inferiority to scientific thought. To a criticism of this kind, the reply is that Shakespeare does in fact communicate Macbeth's sense of immeasurable guilt. He does so by setting up a context wherein the analogy just indicated undeniably communicates his meaning, the same context in which Lady Macbeth's sensible, even positivistic, belief that "A little water clears us of this deed" is pathetically ridiculous. In fact, Macbeth's bloody hand related to the basin of water is the *contextual equivalent* of the guilty soul related to the sea of his life or to his world. The looseness of the analogy is merely apparent. Poetic imagination is manifested, not in using words in an odd fashion, but in creating a context wherein the unusual use of words communicates exactly the subtleties which are to be expressed.

One might say that any sort of similarity between two objects is a measure of the one by the other in respect of that similarity. Such a usage, however, usually is judged by scientific thinkers to be too broad to be of much use. Commonly today the notion of measurement is restricted to a quite specialized context, to the context of quantity. That is, the class of objects usually taken to be the *mensurans* is numbers. Measurement, then, is defined as the construction of a class of numbers which corresponds to, or is isomorphic with, the class of objects to be measured in respect to some quantifiable property. Then, the number corresponding to the object is its measure. The definition may be expressed even more simply: measurement is the assignment of numbers to things in accordance with a rule. Probably anyone today whose subjective powers have been disciplined to a quantitative rationality would accept this definition as unquestionably correct or at least on the right track.

Two things are accomplished by application of this definition. First, the things vaguely apprehended in respect to their quantity—the given presentations—are explicitly brought into the context of quantity. Second, these quantifiable things are related to numbers by means of a technique. Galileo, guided by the Renaissance world-symbols, suggested that the student of nature should measure all that is measurable, and make measurable that which is not. The latter part of this piece of advice suggests the point. If a something cannot be got into the context of quantity, it cannot be measured in the narrower sense at all. Can Macbeth's guilt or sense of guilt be thus measured? Hardly. To understand the special role or art of quantitative measurement, we must understand whether and how a thing can be got into the context of quantity; what the meanings of numbers may be; and how numbers are matched with and give meanings to the objects measured. Thus, we approach the question: are the world-symbols of the Renaissance, which guided men so efficiently in the making of measurements, a trustworthy guide to all the kinds of truth relevant to human life? The fact that the theological character of Renaissance world-symbols was rejected in modern times indicates that the sciences and techniques came to be practiced for new reasons. These new values and motives betray even more evidently than the old the conviction that there are no limits upon the ability of mathematical languages to translate experience into precise and controllable form. The problem, nevertheless, remains: can an X be translated into a number language *without significant loss?* To answer this question, we need to know in some detail just what can be translated into the number languages. To this end the languages of measurement must be examined in greater detail. First, however, a few comments on the instruments and concepts which apply these languages and facilitate bringing an object into a quantitative context will be offered.

## §30. *The Context of Science: Measurement and Theory*

Even before Galileo had given his famous bit of advice, philosophical views had been developed for understanding unmeasured things in a way which justified their quantitative measure. And even before these philosophical views had been expressed, instruments had already been put to work to measure not only the obviously measurable but also that which was not so obviously measurable. The success of these efforts was eventually so great that it could be said the future of physics lay in or beyond the fourth decimal place of accurate laboratory measurements. The determination of Planck's constant, of the diameter and weight of the electron, and of the values of the packing fraction attest to the correctness of this remark. Thus, the destiny of the modern world has turned upon the exactitude of measurement. Little wonder that many of the leading minds became fascinated with the possibilities of precision in these techniques and seemed in the beginning to expect to be able to refine the techniques without limit.[3]

Special experience has traditionally been distinguished by the fact that special instruments have had to be used in order to gather the relevant data, data which find their meaning within the highly specialized and formalized interpretative principles of scientific theory. These instruments serve to bring that which

is to be measured into the context of quantity and to exclude
other contexts. Understanding of some general properties of
these instruments and theory should, therefore, impart a grasp
of the special experience pursued by the expert in measuring.

All arts employ instruments—the axe, the pen, the plow, the
violin . . . But the sciences, arts of acquiring knowledge of
nature, are said to be distinguished by their use of instruments
which extend our powers very markedly, even radically. Some of
these instruments are indispensable for gathering the data to be
used by the scientist. They are often divided into two classes,
scopic and metric instruments. The scopic instruments, e.g.,
the microscope, the óscilloscope, extend the bounds of perception, and unable one to get closer to the very small, the very
large, the very remote. They render perceptible that which
without them would be imperceptible or only vaguely perceptible. The metric instruments, e.g., the thermometer, the
mass spectrometer, increase the precision of perception and
actually mediate the translation of perceptions into numbers.
This distinction, however, is none too sharp. The mass spectrometer also renders perceptible what otherwise could not be
perceived, and the oscilloscope is often equipped with a
measuring device. Also this distinction among instruments is
not exhaustive; the computer is a scientific instrument, but it
appears to extend or to refine the function of calculative
thought rather than the organs of perception.

At what point does a common tool—an axe, a pen—become
a scientific instrument? No doubt the question asks only for a
definitional convention. An approach to such a convention may
be precisely the tool's relation to the context of quantitative
measurement. A tool becomes a scientific instrument or a
scientifically usable instrument if it quantifies or facilitates the
quantification of experience or experienced objects. Scopic
instruments prepare and facilitate this quantification; metric
instruments are used to perform it.

The simple instruments which I have in mind are instruments
like the axmuth circle, the meter stick, the clock, the thermometer, the balance. Their use, like the use of more com-

plicated instruments, immediately brings the object on which they are used into the context of quantity. Other properties of objects meaningful in other contexts are left behind or ignored. Or, as the point may also be expressed, the results of the use of these instruments may be described in one of the four number languages. I offer one example: a condemned prisoner on a pirate ship, doing as he is ordered, climbs on a plank, orients himself 90 degrees away from the ship's heading, and takes five steps, the fifth of which lands him in the "deep six." The quantitative account of this event is limited to reference to his orientation and the length of his walk. A more extended account might mention such matters as his weight or his velocity upon contact with the water beneath. But however far such an account is extended, the part of it which aroused the anxiety of the prisoner and the excitement of his captors, parts which belong to the way in which the situation was experienced, must be omitted.[4] Quantitative instruments do not mediate these "interesting" parts; quantitative language can not translate these non-formalizable factors. But neither do these "interesting" aspects interest the scientist. In the tradition stemming from Descartes, these aspects would be stigmatized as subjective insofar as they are nonquantifiable, and ignored.

Metric instruments quantify and thus formalize the object or experience of it and so bring the object within the context of scientific study, but instruments are not alone in this function; scientific concepts and theory work closely along with instruments in forming this context. Measurement which produces data and theory which uses it, however, are so intimately interrelated that it is difficult to separate them. It is not surprising that physics has been defined as the theory of measurement.[5] Measurements may suggest and confirm theories, and theories may define new objects and suggest new instruments for measuring them. Perhaps theories could be called the instrument of instruments, for they render the production of instruments possible and their use rational. But theories are said to differ from measurements in that their function is explanatory. This difference need not be allowed to conceal a similarity in form. The meaning of explanation was developed in section 18 and will merely be recalled here. The point was made that the correspondence of a numerical ratio (e.g., 1: 7) to the ratio of physical lengths (e.g., one foot of a table length to the whole length) explains the latter or exhibits its quantitative meaning. But such a correspondence is not different in kind from the correspondence of theory to its date. [6]

It is convenient, however, to think of the operations of measurement as entering into the data collection phase of scientific investigation or else into the verification phase, whereas the term "explanation" is usually reserved for naming the relation of a rather complex concept or theory to the objects or data which instantiate this theory (i.e., are explained by it). Thus, a theory of electron behavior in the outer orbits of nuclei explains the phenomenon of chemical valence. Alternatively, the theory is true of the phenomena, or it translates the invariant which occurs in the phenomenal objects of the science into a language intelligible to us. We also recognize that a measurement, if correctly carried out, is true of that which it measures; it translates the invariant recurring in the object into the language of number. Thus, the relation of *mensurans* to *mensurandum* is formally the same as the relation of theory to the objects or data which it explains. We may, then, take the *mensurans* and the 'mensurandum on one hand and the theory together with that which it explains on the other as similar. The structure of their equivalence may be simply exhibited by the following expressions: $X(A,B,C . . ) = Y(a,b,c . . )$ where X and Y are two logically similar ordering relations, and A.B.C . . . are either theoretical statements or numbers, and a.b.c . . . are either factual statement or measured objects. The equality sign asserts the formal sameness or isomorphism of the two, whether on one hand they be theory and facts, or on the other numbers and measured objects. A theory, then, is something like a complex measurement. Still, though similar, a measurement and theory are not identical. A measurement is a numerical datum yielded by an object in response to certain operations; a theory is an explanation of a set of data.

A fully developed scientific theory is an isomorphism which holds generally between a complex system of theoretical propositions and objects at least some of which are either directly or indirectly observable. If the theory is held to be true, then the isomorphism has been shown within the limits of experimental error to have held in the past and may be predicted with some probability to hold in the future. Measurements provide the grounds or evidence from which this probability is calculated. If the predicted measurements are confirmed, then the theory may come to be regarded as an acceptable explanation of real empirical objects. Indeed the reality of physical objects comes to be a function of scientific theory. This last is an important point and needs to be elucidated. I elucidate it by noting briefly a typical instance and by showing the evaluative role of such instances.

A ready illustration is provided by the discovery of the electron. A number of investigations appeared to point to the existence or empirical reality of a very small, perhaps minimal, discrete charge of negative electricity. The question arose among physicists whether this charge or particle could be said to exist or to be real. It was argued that the electron could be regarded as physically real if it could be shown to be a material object, and it would be recognized to be material if it could be seen to obey the Newtonian laws, e.g., the law of entropy, which states that a force impressed upon a mass accelerates its motion proportionately to the force. It was, therefore, argued that if the electron had mass, it would show determinable resistance to a change of force impressed upon it; then if it indicated such resistance in fact, it could be concluded to be a real physical object. A situation was thereupon devised for varying the force impressed upon electrons and then measuring the resultant changes. A stream of electrons was directed into a strong magnetic field. If the electrons possessed mass, then they ought to be deflected by the magnetic force to an extent proportional to the strength of the field. And, in fact, the stream of electrons was so deflected. Predictions of the deflection were measurably verified. The electron was thus established as a real physical object, possessing a small but determinate mass. Also it so happened that the apparatus for identifying the electron was later refined into an instrument for measuring the energy of electrons, the mass spectrometer.

The electron eventually took its place among established physical objects as a real object possessing measurable properties such as charge, size, velocity momentum, as well as mass. By this remark I mean that the concept of the electron became normative. It was numbered among the standard objects to which experience of the physicist was expected to conform. It entered into the formation of the context of atomic physics. Otherwise expressed, this standard object became an interpretative principle and an instrument for further research. Chemistry, for example, was reinterpreted as the study of the behavior of the electrons located in the outer shells or orbits of the various elemental nuclei. Also streams of high energy electrons were directed upon the inner shells of nuclei in order to provide data for confirming or disconfirming theories concerning inner atomic and nuclei structure. It is a foregone conclusion that such interpretations must be limited by or conform to the properties of the electron object. Whatever is discovered must be coherent with the existence of this new standard object, the electron. Should some physical event be encountered which seems not to be thus coherent with the conviction of the objective existence of electrons, it would tend to be received with incredulity; although, the more flexibly minded scientist might see in such an apparent counter instance a possible provocation for new and original work. This physical object even becomes normative for the experience of other kinds of objects; ordinarily experienced objects tend to be evaluated by the sophisticated person through the medium of this object established in the context of physical science. Hence Sir Arthur Eddington described the "real table," in contrast to the merely experienced table, as a cloud of electrons and other small particles. Further and in general, the properties of objects determined by measurement and the scientific object thereupon developed are often extrapolated from the scientific context and regarded as the real properties of real objects everywhere. Specialized laboratory experience comes, thus, to be regarded as offering support for a general metaphysical theory. In fact, the "working philosophy" of the average scientific and technological worker has just this source. We shall return later to this matter of the extrapolation of the scientific object to other contexts. For the moment it will be profitable to examine the languages of measurement within the quantitative context.

## §31. *The Languages of Measurement*

We accept, then, the narrow definition of measurement as the assignment of numbers to things according to a rule. We need first to determine the meanings of the numbers thus assigned. Thereupon we may determine which object-meanings these numbers are competent to elicit. The latter problem, concerning the object-properties translatable into numbers, can be settled partly by reference to the instruments of measure which mediate the translation and partly by reference to the power of the number language in question to translate the given property or properties.

We turn to the number languages and begin by noting that numbers are ambiguous. In this section I shall indicate four senses or languages of number, languages which are sometimes confused or ignored even by those who are constant practitioners of the arts of measurement. A fifth sense of number will be distinguished in the next section. The four more usual meanings of numbers are specifiable by rules which determine their properties upon each occasion of use and hence also limit the kind of object properties to which such numbers are assignable. We consider the four number languages in the order of their complexity.

a) The simplest assignment of numbers to things is illustrated by a baseball team whose players are assigned different numbers. The jerseys carrying the numbers might be said to be the instrument of measure. The rule assigning the numbers is this: each player shall have one and only one number; no two players shall have the same numbers. These numerals in effect function rather like names for the players. Hence, this kind of measurement is called nominal measure. Measure in this or in any other sense is clearly an extension of the processes of interpretation. It is a further reflective interpretation of certain things or properties of things which have already been identified in the earlier phases of interpretation. It must be observed that nominal measurement utilizes no other property of numbers than their distinctness from each other; also no other property of the things measured is denoted than their distinctness. The invariant in this instance, which is the same in both the numbers and the things, is merely distinctness. If numbers — say 1, 2, 3, 4 — were assigned in this manner to the strings on a musical instrument, no information would be given thereby concerning which notes are high and which are low. In other words, so long as the rule of assignment is observed, any number whatsoever may be assigned a given object. The numbers 8, 6, 5, 9 would serve as well as 1, 2, 3, 4 for distinguishing four different strings or four baseball players.

b) A second number language is formed by explicitly adding other properties to nominal numbers which enable them to translate additional characteristics of objects. Thus, a second form of measurement brings into play the ordinal or serial properties of numbers in addition to distinctness. The relation "larger than," for example, will serve to place numbers into serial order. The logical properties — irreflexiveness, connectedness, transitivity — give rise to three rules of serial order. So let x, y, z be numbers, and let them be related by the relation "larger than." Then, (a) no number is larger than itself; (b) of any two non-equal numbers, x is larger than y, or y is larger than x; (c) if x is larger than y and y is larger than z, then x is larger than z. Thus, 4, 3, 2, 1 form a series, since substituting them for the variables in the above rules yields true statements. "Smaller than" similarly is a series-forming relation.

Now certain items of experience, other than numbers thus defined, manifest properties and relations which enable us to place them also in serial or ordinal order. This is to say that the relation ordering these non-numerical things exhibits the same logical properties (being irreflexive, connected, and transitive) as the serial relation ordering ordinal numbers. Such relations are logical equivalents of each other. For instance a collection of stars and the relation "brighter than," a collection of stones and the relation "harder than," a collection of strings and the relation "longer than," all are series-forming; trial will show that all three of these relations have the same logical properties as the serial ordering relation "larger than." Let us try "longer than" as it describes the relation of the strings on a Greek lyre. It must be assumed that the eye can already perceive which of two strings is the longer.

The Greek lyre had four strings tuned to four different notes. Hence, if the tension of the strings were equal, the lengths of the four strings would be different. Now, of four such strings any two would be different in length; and one out of any three strings would be longer than the other two. If the longest string is laid down first, the three ordinal rules determine the order of strings to follow, just as earlier these rules determined the order of numbers, 4, 3, 2, 1. The relation "longer than" among strings and the relation "larger than" among numbers possess the same logical properties; they are logical equivalents, however different they may be in other ways. Hence, in this respect, the two may be matched with each other; that is, the series of numbers may be paired off with the series of strings. Here we not only match a string with a number, showing that the string like the number is distinct from the others in its class; we match up the serial properties of numbers and strings, so that the numbers translate the place of the strings in a serial order. When this kind of pairing relationship holds between two classes of objects, it is said that the two are isomorphic with each other; they share the same form or invariant structure. Then one of them measures or interprets the other.

When serial order among objects is important, it is often useful to select a handy group of objects placed in numbered serial order and to use it as typical or standard for such objects. But in order to use this standard set as a scale, equality must be defined. Hence, we stipulate an object X in the standard set is the equivalent of some other object A with respect to some serial relation, say "brighter than" if and only if X is not brighter than A and A is not brighter than X. Then, if X and A are stars, they have equal brightness or magnitude. If X is assigned a number of the standard series, then A may also be assigned the number of X; A is thereby measured. The Beauford wind-force scale, the Mohrs hardness scale, scales for measuring intelligence are other examples of ordinal measurement.

It is always important to recall that the properties of numbers on an ordinal scale are just those properties which the ordinal or serial rules can translate. If as above we were to number the strings on the Greek lyre, 4, 3, 2, 1, these numbers would order the strings in respect to their length. The numbers in this usage are isomorphic only with the order of tones with respect to lower and higher. They say absolutely nothing about how much higher one tone is than another; much less do they serve to compare musical intervals. In order that numbers reflect these other relationships among notes, additional properties and relations must be explicitly assigned to the measuring numbers. Numbers are the disciplined servants of the measurer; they translate only just those properties which the rules of their language allow.

c) A third language can be generated by defining equal intervals between the elements of a number series. These affine numbers are described by the function $y = ax \dagger b$, where a and b are constants and y is a variable dependent upon x. Equal intervals can also be defined for certain kinds of experienced objects. In the general expression for this kind of number the "b" is an arbitrary constant and suggests that the beginning point of a number scale of this sort is arbitrary (i.e., convenient). Notes on a piano, for example, are so defined. If a whole tone be regarded as the unit interval and if the scale be begun at middle C, then F is 2 1/2 notes higher than C and G is 3 1/2 notes higher, and so on. Another familiar example is the scale for measuring heat intensity. The Fahrenheit scale sets its zero point at the freezing temperature of salt water and its upper point at the body temperature of man. The range between is divided into 96 equal intervals. On this scale pure water freezes at 32° and boils at 212°. Clearly the zero point could have been placed elsewhere and the unit degree could have been defined differently.

Again, it is important to observe that these numbers interpret only the invariant which they have been constructed to interpret. In the case of the thermometer, the numbers refer only to heat intensity, not to quantity of heat. A glass of water at 50° and a piece of aluminum of the same size or of the same weight do not necessarily contain the same quantity of heat. No more is B flat on the piano twice as high as F, even though B flat is five notes above middle C and F is 2 1/2 notes above it. This point is obvious when one recalls that the beginning point is arbitrary. Zero is not a given operator in this number language. A zero or beginning point can be set *ad libitum.* Also multiplication is not possible in it. Thus, numbers on these equal-interval scales do not translate ratios or products but only the number of equal intervals. In order to express and to translate equal ratios, numbers have to be endowed with additional properties. I illustrate the character and use of these more richly endowed (cardinal) numbers in a fourth kind of measurement by reference to an early and momentous instance of their use.

d) It will be observed that the four tones on the Greek lyre, in the Doric mode, make two perfect fourths when struck. That is, the first two notes make a musical interval which the ear recognizes to be the same interval produced by the second two. In modern notation they are: . The ear has to detect the fact that the relation or interval of C to F sounds exactly like the relation of G to C'. We want now to translate the sensed equality of the two sound ratios into ratios of numbers. In this instance the number language must translate equality (or inequality) of ratios.

In order to understand the notion of comparison of ratios among numbers, each number must be thought of as a collection of units, i.e., as containing the unit number a definite number of times. Thus the number named '6' contains the unit six times, or as it might be expressed, $6/1 = 6$ (here the slant means "is related to"). Thus, the cardinal number 6 is a brief expression for $6/1$. Then the interval from 6 to 8 might be expressed as the ratio of $6/1$ to $8/1$ or $6/1/8/1$, or more briefly $6/8$. Cardinal numbers, including zero, have nominal, serial, and equal-interval properties; also they permit the operations of addition and multiplication and their inverses.[7] These are the numbers which are used in many of the daily affairs of the market place. They are the ones Pythagoras used about twenty six centuries ago to perform the measurement of the lyre in an experiment which might be said to have initiated mathematical physics and to have heralded from afar the age of technology.

Pythagoras used a monochord, a gadget having a string under constant tension and equipped with a moveable bridge, so that the tone emitted by the vibrating string was a function only of its length. Four similar monochords were adjusted so that the four notes of the lyre C, F, G, C' were sounded. It will be recalled that the ear recognizes the sameness or equality of the interval C/F to the interval G/C'. If, now, the shortest string (C') be taken as the unit length, then the lowest note (C) is easily seen on the monochord to be twice the length of this unit, or C'/C as $1/2$. Then, also, F measured by this unit, is 1 1/3 units in length, and G is 1 1/2 units. Thus C', F, G, C are respectively, 1, 1 1/3, 1 1/2, 2 or 1, 4/3, 3/2, 2 units in length. Do these numbers reflect or translate the equal ratios of sound? If they do, then $1 : 4/3 = 3/2 : 2$ would be a true statement, and would reflect the equality of the musical intervals. Of course the relationship among numbers expressed in this manner is difficult to see; but if these numbers be translated into whole numbers ("multiplied through" by six) so that the fractions disappear though the ratios remain the same, we get $6 : 8 = 9 : 12$. Here the obscure ratios are expressed in a form where the equality or inequality is evident. It is in fact true that 12 is related to 9 by the same ratio as 8 is related to 6; the test, that the product of the means should equal the product of the extremes (Euclid, Book V), demonstrates it. Or otherwise tested, $12/9 = 4/3$ and $8/6 = 4/3$, and $4/3 = 4/3$. Here the subjectivity trained to quantitative rationality can most easily "see" that the numerical ratios or intervals are equal. We may, therefore, conclude that the equal ratios of numbers interpret or measure the ratios of the equal intervals of sound. The numbers and the strings are in this regard isomorphic. Here the equality of the ratios is the new invariant or form which is translated by cardinal or complete measure from the sounds to the numbers, where it can be intellectually "seen." Only the equality of ratios, however, is thus translated from the lyre into numbers. The content of the music, its history, its effect escape translation. It is of the utmost importance to note that the meanings assigned to cardinal numbers precisely limit their possibilities of translation. Such must be our conclusion.

It, nevertheless, is easy for us to understand why the Pythagoreans, following upon the dramatic discovery of the numerical essence of the lyre, felt that they had come upon the secret of the universe and of man. They quickly began discovering numerical essences everywhere. Plato was fired with the same enthusiasm but reined it in with metaphysical restraint and subjected it to a more mature wisdom. Renaissance and modern scientists returned, though with vastly more patience and appreciation of difficulties, to the same Pythagorean *élan*. It must be added, however, that the knowledge thus acquired has come in the Western world to be evaluated and used in a manner quite different from that envisaged by Plato and perhaps by the Pythagoreans. In the Cartesian context the aim of the mathematical understanding of the cosmos is no longer the perfection of man by introducing the harmonies of the cosmos into the mind and self of man; rather, the end and aim comes to be to increase man's power by making him the "master and possessor of nature." Here the technological possibilities hidden in the Pythagorean measurement of the lyre begin to manifest their unbounded scope.

A brief summary of this section will be useful for emphasizing the characteristics of measurement which the modern inheritors of Pythagoras are tempted to ignore. Measurement has been treated as an interpretative technique by means of which the

quantifiable properties of objects are translated into numerical
languages and enter thereby into scientific theories and finally
become subject to technological control. At the same time
certain limitations inherent in this technique of translation were
exhibited. Numerical languages are languages of quantity and
are capable of translating only the quantitative properties of
objects. But we observed that the notion of quantity is complex.
A property is quantitative if it has the first of the following
hierarchy of characteristics or any successive ones: it is constituted of enduring and distinct identities (though what these
identities are is not further specified); these distinct identities
may be placed in serial order; equal intervals may be defined
among them; finally, ratios may be specified among these
intervals which can be shown to be equal or unequal.
Measurement understood as the assignment of numbers to
things according to a rule is limited to statements within these
bounds. Although measurement is a very powerful instrument,
it is not all-powerful. It cannot translate other properties than
quantitative ones; as it were, it has no names for any other
properties.

## §32. *Genuine and Questionable Measurement*

It is important to note that imaginable variations in the ways
of measuring the lyre yield the same ratios. Suppose the strings
in equal tension were measured by a foot rule. If the measuring
rod were one foot long, then the measurements of the four
strings in inches might be: 24", 18", 16", 12"; for these
numbers are reducible to the ratios already indicated. Also the
strings might have been of one length and the tension varied
and measured. Or one might count sound waves as they pass a
given point. All these measurements would yield the same
ratios. P.W. Bridgman held that the test of genuine as opposed
to pseudo measurement lay exactly in the fact of several
procedures of measurement leading to the same result.[8] Such
caution is necessary. Although measurement is an indispensable
key to the secrets of nature, still it is a key which can be misused.

Perhaps the most egregious kind of misinterpretation arises from ignoring the particular number language, whether nominal, ordinal, equal-interval, or complete into which the *mensurandum* is being translated. Thus, numbers used to measure the order of a set of elements with respect to some characteristic (and which, therefore, have only serial properties) are nevertheless sometimes multiplied, divided, their average found, etc., just as if they were cardinal numbers. Some social scientists have even been heard to aver that their journals publish some seriously offered pseudo-measurements of this sort.[9] The source of the error on the formal side is obvious; it lies in the fact that the different senses of number—nominal, ordinal, equal-interval, cardinal—are all indicated by the same symbols, i.e., numerals. And the unsophisticated users of these symbols jump to the conclusion that the same symbol always has the same numerical meaning. In fact, though the same numerals may be used in different number languages, the meaning of the numerals is given not by the mark on paper but by the rules of the language in question.

It is equally essential to recall that properties are not assigned objects by assigning numbers to objects. Rather, if measurement is to be objective, the number language must be chosen to accord with the properties which objects already have within the given context. As is obvious, space of today's common sense can be divided and measured in each of the four ways discussed above. No doubt the division of the walking space between two places into pace-lengths and counting unit paces is an ancient technique. It differs from modern procedures only in that the unit length and the operations of applying it are more precisely defined. But in either instance, space must be prepared for measurement by being emptied, so to speak, of non-quantitative, experiental, metaphorical, and mythical content. Nothing remains after this preparation but spatial quantity understood as structured in one of the four senses which we have designated. That space *ought* always to be so prepared, that it is better or truer so, is a Cartesian metaphysical postulate aptly named by a philosopher "the axiom of impoverished reality."[10]

Time likewise can be subjected to all four kinds of measure. Distinct times may be given numerical names as if they were baseball players. Temporal points may be judged earlier-than or later-than and assigned numbered positions in such a rank-order scale as a calendar. Equal intervals of abstract time may be defined by (say) a pulse beat or by the diurnal motion of the sun. Instruments, such as the clepsydra or the clock, may be devised for counting and recording these equal measures. Also equality of ratios or temporal measures or temporal-spatial measures may be discovered; for example, when I drive my car at 30 miles per hour, I am illustrating the general proposition that the unit of distance is related to the distance per unit of time as the total time for the trip is to the whole distance covered (i.e., $1 : r :: t : d$, or $d = rt$). In other words, time and space, when quantified, contain among their structures invariants which may be translated by the appropriate quantitative languages. And like space, time understood thus tends in the modern age to be given a metaphysical privilege.

Object properties, similarly, can to various extents be translated into the several number languages. Instances of redness, for example, can be nominally measured. Also samples of redness may be placed into a series ordered by the relation "more red in hue than." I know, however, of no equal-interval scale for measuring redness. Can redness be given full or cardinal measurement? At least redness can be so measured if it can be correlated with its temporal and spatial properties. And in fact, electromagnetic waves of a certain length from crest to crest, moving with a definite and measurable frequency per unit of time can be correlated with the perception of different hues of redness. In this indirect manner the spatio-temporal characteristics of redness can be interpreted by cardinal numbers. The numbers thus assigned become a kind of metaphor for the non-quantifiable properties with which they are correlated.

Many other object-properties inaccessible to direct measurement can be reached by such indirect methods. Indirect measurement is usually defined as the ratio, or some other function, of two or more quantities which are directly measurable. Velocity offers a familiar instance: it is the ratio of the distance a body moves to the clock time it takes to complete the movement. It is taken as obvious that space and time are directly measurable. The attempt, however, to translate all the discernible properties of the world and of man into numbers, whether directly or indirectly, is sometimes disappointed. Pseudo-measurement also exists.

The difference between genuine and pseudo-measurement turns upon the understanding of two points. First, the category of quantity must be understood to be more complex than an older philosophy believed it to be. I have described the *mensurans* as constructed of four hierarchically organized languages whose grammatical rules are clearly expressed and need not and should not be confused.[11] Secondly, the *mensurandum* must be shown to possess the properties correlative to the number language into which it is to be translated. That is, the structure of the *mensurandum* must be seen to be logically similar to the logical structure of the *mensurans*. The assignment of nominal numbers to flames of fire, of ordinal numbers to instances of feminine beauty, of equal intervals to lived events in time, of cardinal numbers to intelligence, are all instances of pseudo measurement. They are inauthentic imitations of scientific precision.

The meaning of "assignment" should by now have become clearer. We speak of assigning properties to numbers in order to form different number languages; we also speak of assigning numbers to things in order to perform the corresponding kind of measurement. The sense of "assignment" differs in the two usages. The first tense of "assignment" is syntactic; it determines the structure of the number language in question. The second sense is semantic, the dictionary sense. It determines the extra-numerical meaning or reference of the number language. In order that this semantic reference be possible and measurement be genuine, the *mensurandum* must be shown to exemplify a similar syntax to, or to be logically homogeneous with, the number language selected. For the present non-formal purposes, this second sense of "assignment" is, when genuine, the interesting one.[12] for it is this which enables a number language to convey information about something non-numerical or not exclusively numerical. It is the means by which the old knowledge of numbers becomes new knowledge of experience. Recognition of this power of the mathematical languages was the epoch-making discovery of Pythagoras and the insight which convinced Galileo that he at last possessed the key to translating the "book of nature."

These observations on measurement serve also to bring forward a more general principle of the greatest importance, *viz.*, that a close and regular relation exists between a language and that which can be said in that language. The elements and relations composing the language must be at least rich enough to express what is to be said; conversely, that which is to be said must manifest a complexity commensurate with the language. In other words, the language and that to which it is to be referred must be logically similar. This prescription is a necessary condition for the literal use of language. The possibility of a literal use of language about an object is the condition for rendering the object scientifically intelligible. The advantage offered by the study of measurement is that this principle of literalness both in its genuine and its questionable expressions, is exhibited in a relatively well understood language, applied in a disciplined manner to simplified (or abstract) situations. This usage approaches the ideal of precision which is determinative for the sciences. Whenever a mathematical language is insufficiently rich to express what its user desires to express, then new terms, definitions, axioms, or operations may explicitly be added. And if these additions genuinely reflect properties of the objects referred to, then the literal usage may be continued. Where, however, that which is to be said is too complex, too subtle, or too tenuous to be expressed in language used literally, then the speaker must resort to non-scientific techniques or else fail to communicate.

### §33. *A Fifth Sense of Number*

Some philosophers have attempted to extend the ideals and techniques of measurement and of the sciences to all thought and to all language use. This attempt is the natural consequence of their singleminded devotion to procedures similar to those described in the preceding two sections and to the metaphysical principles characteristic of the Seventeenth Century, principles which were intended to justify the universal application of these procedures. These principles, however, have undergone an important change since they were developed by Renaissance man. They have been demythologized, secularized. (This change will be considered in the next chapter.) The later positivistic and naturalistic view is often expressed in terms of the sufficiency of the language of physics and chemistry, mathematical languages mostly, for all acceptable thinking. Expressed more generally: clarity and precision is the ideal, but "to clarify is to formalize," or "to be is to be the value of a variable." From this standpoint "the axiom of impoverished reality," mentioned in the preceding section, seems to follow as a necessary consequence. This "axiom" is, in fact, a metaphysical inference which expresses in brief an important part of the cultural belief characteristic of the era of Galileo and Descartes. Here we shall have to raise the question again whether language used literally and the mathematical method of enriching language are sufficient for describing and discussing everything that exists. Or is there need for other methods of enriching language?

The conviction that mathematical techniques for enriching language are sufficient for all uses is equivalent to the conviction that there are no differences in principle between the special languages of the sciences and ordinary language, no differences in principle between special and common experience. Positivism holds to this methodological monism. It interprets ordinary language uses simply as a confused and inexact illustration of that which the scientist does with skill. It believes common experience to be merely special experience misunderstood or vaguely understood. Conversely, science is refined common sense. If this view were true, then the humanistic use of language is merely inexact science; then Macbeth's remark concerning the incarnadined sea (cf. p. 162) could have been better — because more precisely — expressed by a physicist using an interferometer. This, however, is ridiculous.

I pointed out (*ibid.*) that Macbeth's remark is already exact for it exactly translates into English the transforming recognition of monstrous guilt. If it be asked how this conclusion was reached, the answer will eventually have to be analogous to a point made earlier, *viz.*, that the measurement of equal intervals of sound depends upon the ear recognizing which intervals are equal, (or upon the eye seeing that certain pointers register the same). And then one must "see" that the equality of intervals is reflected in the equality of numerical ratios. Analogously, one must "know" or empathetically experience guilt to see that Macbeth's expression translates it. But Shakespeare's expression is a metaphor. The "knowledge" it conveys is not quantity, not translatable into any one of the four numerical languages. Although an experimentalist might speak of Macbeth's blood pressure or skin temperature as bearing some exact or measurable relation to those of Lady Macbeth, it is manifest nonsense to assign numbers to guilt and to conclude, for example, that Macbeth is literally twice as guilty or one third as guilty as his wife. Plato understood this nonsense better than some later measurers when he solemnly demonstrated that the just man was exactly 147 times more happy than the unjust man. In short, the mathematical methods for the enrichment of language do not suffice for all needs. The objects about which we speak have or can acquire non-quantifiable meanings.

Some ordinary language usage, the language of the poets, scholars, and some philosophers, *viz.*, those preoccupied with understanding or enlarging common experience, rely upon other methods for enriching language. Among the ways of elaborating new meanings and terms is the analogical and metaphorical development upon the basis of old ones. A few remarks upon the analogical and metaphorical means of enlarging language will help to indicate the limits of quantitative methods.

The formal structure of analogy, A : B as C : D, resembles a mathematical statement, except that the relation, ": ", is not equivalent in the two instances, only similar. Also, it is not defined, only rendered evident, if implicitly so, by the context. Still, one pair of the related terms interprets the other pair with respect to this similarity. The similarity may be quite subtle; usually it depends upon the context to elicit the exact interpretation appropriate. Consider Shakespeare's description of sleep from *Macbeth:*

"Sleep, that knits up the raveled sleeve of care." Here the analogy may be expressed in a more pedestrian fashion thus: sleep : **care worn** body as knitting : ravelled sleeve. The intended peaceful and restorative quality of sleep is evoked by the just suggested image of the care-worn laborer, to which our imaginations are ready to add the image of a woman sitting by his bedside quietly, mending the frayed sleeve of his work jacket. This is the sleep which Macbeth might have enjoyed but can no longer. It is Macbeth's self which cannot thus be restored; this he intimates a few lines further.

To know my deed 'twere best not know myself.

In addition, and for purposes of saying that which literal language cannot accomplish, metaphor is sometimes a more effective means than analogy. According to Aristotle (*Poetics,* 21), metaphor consists in giving one thing the name which belongs to another. The basis for this exchange of names is again some similarity among the things named. In the happier instances this similarity leads the mind to see something not previously seen or not seen so well. Metaphors may be understood to be generated from analogies, usually by suppressing some part or parts of the analogy. Consider again in *Macbeth,* a line from the protagonist's response to the announcement of his wife's suicide: "Out, out brief candle!" The originating analogy is something like this: as a candle is to the room it illuminates, so her life was to his world. Thus, the familiar candle conveys something of the elusive and strange sense of this life. Both candle and life counter a darkness, as both may flicker out into a night which signifies nothing. King Duncan had given light and structure to this world. After the King was murdered, there remained only brief candles which flickered out to leave Macbeth facing self-destruction. Here as in the earlier mentioned metaphor, the one which takes the tumultous and incarnadined sea for Macbeth's blood-stained soul, the imagination, provoked to an extreme by fear and guilt, leaps to a new and fitter meaning by this exchange of names.

In general, the pursuit of the non-objective dialectic interpreting the modulations of self-experience, quickly moves beyond the range of words used literally and strains the resources of even the richest and most supple of languages. The poetic and dramatic employment of language are no less necessary for pursuing this dialectic than the languages and techniques of measurement are for executing the tasks of science and technology. The one is as admirably fitted for pursuing the non-objective dialectic as the other is for the objective.

The need for enlarging language beyond the level of the literal invades even mathematics. This need is encountered by anyone who seeks the meaning of (say) the number two used to "count" two concrete individuals. Socrates was the first to note the oddity in the fact that, though he and Cebes are each one, yet if they are juxtaposed, then somehow together they become two (*Phaedo* 96D). In what sense are they two? The ambiguity of numbers, already illustrated, should warn us to be skeptical of such an assignment.

Socrates and Cebes are human individuals. Each such individual is a unique being in a concrete and ontological sense. None can, *as such,* be subsumed under a general concept, unless in a negative way which specifies the inviolability of his uniqueness. It follows that when we speak of two concrete individuals, "two" is not given a literal but a figurative sense. In order to conclude that Socrates and Cebes together form a (quantitative) group of two, the measurer must ignore the Socratic character of Socrates and the Cebean nature of Cebes.

He must abstract from them their concrete and individual being so as to have left just two similar and countable objects. For to count individuals they must be the same in the respect in which they are counted. If the "two" individuals are concrete and unique, they are not literally the same. Then they cannot be counted. Thus, this concrete "two" refers us to unlike component unities. We may call this kind of unit pre-mathematical, for it cannot be used in counting objects but only for referring to objects before abstraction from their unique being has been made. It has sometimes been called a "dialogical number," for a typical use is to refer to persons engaged in dialectical conversation or action whereby they are undergoing change and cannot even be literally said to be in every sense one with themselves, much less one of two or more similar or countable objects. Macbeth, as Shakespeare saw him, was undergoing such a change; he was engaged in the act of discovering the violence he had done himself. A self pursuing the truth about itself and searching out its own unity, is seeking its oneness in this fifth sense of number; he is striving to become his own unique individuality (cf. §35). As such, this self is the contrary of the mass man, of the alienated man, or of a man who is merely the role player or functionary. Each of these latter has identified himself with only a part of his whole human being. Each might, therefore, be said to be a metaphor for the whole human being. We do not, however, ordinarily dispose of terms for the whole of any such being. If every-day language is language used literally, then we must always depend upon metaphor or upon some such figurative device for referring to a complete being or unique unity.

Can this fifth and figurative sense of number be utilized in anything like a process of measurement? Since properties must be common in order to be measured — measurement being the comparison of similar objects in respect to length, degree of temperature, place in a series, or the like, — one would initially be inclined to deny that number in this fifth sense can represent the measure of anything. Nevertheless, comparable or measurable properties must, at least in common parlance,

belong to an object. That is, objects have many properties and
these properties are always unified in a manner which in each
individual instance is in some way unique. This pen, for
example, is unique at the least in its location at this place and at
this time. Perhaps the premathematical sense of number can
refer to just such unique unities. I think it obvious that an
individual human self qualifies as such a unique entity. At the
least he occupies a unique locus as the center of his own lived
time and space. He is a unique one. As such he is not yet defined
as a member of a group of similar individuals, nor is he
measured or measurable by reference to the other four senses of
number. In this negative sense, the pre-mathematical sense of
number might be said to offer a sort of negative measure. A
unique one is one, but it is neither qualitatively nor quantitatively measured. It is one being in its concreteness. [13]

The pre-mathematical unit is involved when the individual in
his individuality is invoked. This invocation often calls upon the
techniques of literature or of fine art to elicit the range of
response adequate to the awareness of a concrete individual (cf.
§11). This invoking Shakespeare achieved in *Macbeth* and
Heidegger, using a quite different method, sought to accomplish in the instance of an object, an earthenware pitcher. [14]
In daily life, no doubt, it is approached in friendship, where the
friend is recognized and valued simply as himself. The experienced uniqueness of the friend is consequent upon uniting
in perception, so far as possible, all the relevant experienced
properties, traits, and event in a manner characteristic of the
friend and of him alone.

In fact, though, even the best of us fails to achieve complete
unity. We of the West still see our completeness in the indefinitely remote future and can render it present and effective
only by means of the world-symbols characteristic of our different cultures (§27). This unity and completeness is envisaged
sometimes as a personal unity with a creator God, sometimes as
a rational unity with a divine creator-intellect, sometimes as a
sense of poetic unity with the cosmos, sometimes as a
cooperative unity of all men engaged with machines in the task

of achieving complete domination and exploitative control over the environment.

The fifth sense of number, thus, can point to the complete but ideal unity of a man as well as to a unique, existing self. In either instance, this figurative unity is useful for referring to a concrete individual in just the sense in which the number languages of the preceding section do not apply. For if a group of such unities is described by some property which all its members share in common, it is obvious that the uniqueness of each escapes this description. Such unities are said to be abstract. A man, for example, is abstractly definable only in the sense in which he does not differ from other men. The tragedy of the sciences of man is that the exact sense in which a man is subject to a science is just that in which he is an abstract man, that in which his uniqueness is ignored. For only when his uniqueness is set aside is he prepared for definition or for measurement, and just then he is regarded only as similar to every other in the group. A group of men, that is, can be counted only after each concrete individual self has been (mentally) replaced by a standard, abstract, and thus countable man.

However useful concepts of individuals having properties in common may be, such concepts can be used in an investigative enterprise only under the shadow of a real or feigned ignorance of the individual's uniqueness. A subjectivity disciplined in the objective dialectic and skilled in the arts of measurement is also a subjectivity which has learned to ignore uniqueness. But ignoring an individual's uniqueness may upon occasion be a major loss. The loss can be disastrous when the ignorance is real. The qualitative measure of a man by reference to a job description, to a definition of some required role, or to some moral ideal belonging to a culture is often necessary, but this measurement is justified and sophisticated only when reparation is made by recalling the uniqueness of the person measured. Otherwise, the abstraction does him violence. By defining man as the individual being obligated to inquire into his own nature and destiny within the limits of the human condition, I intend to include a recollection of the uniqueness of each. Does technological man retain this recollection? This man is an expert in making precise measurements involving number in the four senses of section 31, in developing theories therefrom, and in using these theories to dominate nature and to force it to subserve his desires. Whether he and his culture preserve a lively recollection of the fifth sense of number in its reference to the beings which they measure and seek to control remains to be seen.

# CHAPTER VII

# THE MEASURE OF TECHNOLOGICAL CULTURE

## §34. *The Standard of Measure*

The scientist is trained to make quantitative interpretations of his experience and to reason mathematically upon these data. Thus, his subjectivity is disciplined to the kind of thought and technique indicated in the preceding chapter. The man of technological culture, the topic of the present chapter, goes a step further. His reason is disposed exclusively to this usage. He will tolerate no other.

Have we reached the stage where this latter kind of reason completely dominates our culture? It bears reiterating that this essay does not undertake to decide whether any Western or Westernized society embodies technological culture.[1] Indeed, it is irrelevant to our purpose whether or not any people have exemplified or ever will exemplify a completely technological culture. The concept of such a culture, like that of a frictionless plane, can serve, nevertheless, as a measure. The closer an existing people come in relevant particulars to instantiating the concept of a technological culture, the more pointedly will the judgments to be rendered in this chapter apply to it. For a long time, however, the major energies of Western countries have obviously been directed toward the invention and production of machinery to replace human labor and to adjust man and his environment to the use of these machines. Moreover, these same countries are in process of learning to use new types of machines, including many kinds of computers and cybernetic devices which promise greatly to alter and to accelerate the process of mechanization.[2] Have these countries thereby become "technologized"? The problems concerning the kind of relations which the peoples of these countries will or should establish with machines and what the establishment of these relations will do to the peoples so involved, are continuing and pressing. This attempt to measure a fully technological culture is intended to be a contribution to determining just what the relation of man to a fully mechanized civilization would or should be like.

Were we in possession of an adequate definition of the human essence we might proceed confidently to take the qualitative measure of technological culture by matching this definition with the typical man of this culture and by judging the degree to which the latter fulfills the requirements of the former. Since, however, we possess no such definition and since man's capacity to undergo radical change may be judged to entail that no such definition of him is possible, we shall have to proceed otherwise. We have observed that in manifesting his being a man acts, and his action is limited by the world he inhabits and by the culture to which he is bred. We have taken notice of two worlds: the Classical Greek, with its conviction that a man can and ought to pursue his perfection as defined by his essence, and the Christian with its persuasion that man is nothing of himself and that to pursue his indefinitely remote completeness he needs help from beyond himself. The latter world has developed widely diverse cultures. The occasion has arisen to mention three of them: the Mediaeval, Renaissance, and Modern cultures. To these three a fourth may be added which I shall call "technism," the fully technologized culture. Technism, however, may be something other than still another variation upon Western world cultures.

These several cultures, with the possible exception of

technism, are at one in that all are still sensitive to the persuasive power of Western world-symbols. All see the salvation of individual man as the proper goal of human endeavor. But all see individual man as powerless to save himself, and all place his completeness, the goal of all his effort, at an immeasurable distance from his present state. All combine pessimism regarding the individual unaided man with optimistic progressivism regarding the goal of man in society with God, with enlightened others, or with modern machines. As the later cultures move away in time from Mediaeval civilization, they also become more and more remote from the Christian interpretation of the fall of man and of his salvation; their view of these world events becomes more and more secular. Is there a point at which a culture of this type ceases belonging to the Western world and passes over into a different world? The problem is not merely an academic dispute over names. World and culture (as viewed in this essay) are different in kind. World is the possibility of acquiring a culture (§4d), but it is a limited and structured possibility. World expresses the finiteness of being; if some possibilities come into use, others are thereby rendered unavailable. Not all of the characteristics of Classical Greek and Western Christian worlds are compatible. Despite their many common traits, one cannot be an inhabitant of both worlds. Any two distinct worlds are, in some important respect, incommensurable, and to move from one to the other requires a radical change in a man. But it is possible to become a member of more than one culture within a world. Sir Isaac Newton was both a Christian and a mechanistic scientist. Also, there exist Romantic technologists.

Cultures change and evolve. Is it possible to determine the point at which a culture becomes incompatible with the world in which it originated? Or do human possibilities undergo unaccountable mutations? In any event, when the symbols belonging to a world lose power to inspire and to direct, then the world-structure must be so altered that the culture eventually does pass over into another world. There is reason to suspect that technological culture is now approaching that

point of metamorphosis. Just possibly its later development will
have to be evaluated as a radical change. The event will have to
tell. By describing and measuring technism we shall attempt to
perceive the route of passage into a possible later world.

Our purpose will require use of the brief summaries of the
investigations developed in Chapters II through V, for so far as
these summaries do represent and recall insight into human
being and the human condition, they may be used normatively.
Thus, we proceed by determining whether technological culture
or technological man effectively accept the obligation
courageously to pursue the truth about the self to which
Western world-symbols point and the guiding recognition of his
dependency, finiteness, and non-identity. To this end we have
first to determine what a technological culture is like and
whether or not it invites or at least renders possible the
discharge of the first human obligation, the inquiry into
oneself. And we need to make the same determination with
respect to technological man. In particular, we need to
determine whether this man seeks in principle to discipline his
subjectivity to the kind of reason which can successfully pursue
the non-objective dialectic. Secondly, the way in which the role
of scientist-technologist is envisaged in a technological culture
needs to be studied; that is to say, the parts of the role structure
will have to be examined with respect to the way in which they
preserve recognition of the basic character of the human
condition, its dependence, finiteness and its non-identity with
any role. Thirdly, the qualitative measurement of technological
culture and man thus completed, its results should be sum-
marized in a concluding judgment.

## §35. *Technological Culture and Man's First Obligation*

Although I am not asserting that any Westernized country is
as yet fully technological, it is evident that several of them desire
to become so and are making progress in that direction. Thus,
one has only to look around in order to form a rather good
notion of the nature of this kind of culture. I have merely to

render this notion not more detailed but more exact. In this section I seek to disengage and evaluate an essential property of such a culture. Needless to say, this property has nothing to do with any specific type of machine or usage to which a machine might be put; it has to do rather with a factor in the modern culture which inclines men to be disposed to work in technological contexts, and these contexts are an essential factor in determining how objects, the self, others, and nature are to be interpreted. Technological culture can develop, I want to point out, in such a way that essential world relations are changed, with the consequence that a new world may come into being. And the cultures of this new world may well render the discharge of man's primary obligation (cf. chapter III) all but impossible to accomplish. He may, indeed, harbor no motive to accomplishing it.

What, then, is technology? In a generic sense, technology refers us to any characteristically human use of instruments for any sort of production. So broadly understood, techniques would include primitive agriculture as well as modern applied science. But ordinarily the term is more narrowly used. H. D. Lasswell's definition has become well known: technology is "the ensemble of practices by which one uses available resources in order to achieve values." This definition, however, fails to characterize the whole of what Jacques Ellul has called "the technological phenomenon." For into this phenomenon consciousness and reason explicitly enter.[3] Ellul's view will not be too much distorted, I think, if his "consciousness" be identified as my "felt possibilities" and his "reason" be identified as applied science or measurement and calculation used with an eye for practical efficiency. More explicitly, this reason is subjectivity disciplined by the measuring skills which were indicated in the preceding chapter and by other theoretical and applied quantitative techniques like them. This reason, exhibiting itself as a technology of the machine, began to be intensively cultivated rather early in the modern era as that upon which man might depend to remedy his deficiencies and to complete his nature.

As it might be roughly expressed, a technology of the machine plays the part in modern culture which a theology of Christ and the Church played in Mediaeval culture. The essential first steps toward this modern era were taken by those men of Renaissance genius who conceived of an abstract space, time, and object (§29). Their strategy consisted in replacing the object as it manifests itself to perception in all its qualitative fullness by an abstract object limited to its quantifiable ("primary") properties. In the preceding chapter we described how the quantified object is measured and subjected to the sciences. This replacing of the perceived object by the quantified and measurable object became a metaphysical decision when the quantified object was detached from the fully manifested object and endowed by the Renaissance (and Modern) cultures with an independent and superior reality. Techniques of measurement could then be believed to provide access to the whole of nature. The possibilities inherent in this beginning were seen through to their conclusion by Auguste Comte. Comte elaborated a positive metaphysics of the sciences organized into "integrative levels" of increasing complexity, formulated the famous law of the three stages according to which the most mature stage of the development of a type of inquiry as of a whole culture is the scientific stage, and envisaged the complete "scientizing" of society under the direction and control of politico-scientists. Control over the feeling and thinking of the population was to be maintained by means of a religion of science. Under this religion, individual man was nothing of himself, but the species was deified. And the individual's and society's dependence upon the machine and upon the technological thinking which produced and continually improved the machine was complete. This program proved feasible, and we have in fact realized many of its prescriptions. Only the religion of science proved to be a weak link. Nevertheless, Comte's intuition was accurate; if a culture is to become through and through scientific and technological, it and its world-symbols must acquire something like a religious dimension. Indeed, the evidence indicates that as any culture develops, its goals will acquire a religious aura. But this quality is not to be sought in some manufactured doctrine and practice consciously glorifying (say) technology; rather, it is to be sought in the spontaneously and commonly felt valuations which immediately and "naturally" incline members of a culture to pursue goals in common. There is evidence that presently in the West a rather mystical valuation of technology is indeed felt as the motive to acquiring this kind of culture.

The goal pursued, of course, is complex. Most observers would agree, I believe, that the distinctive values unquestioningly accepted by modern Western cultures through the Nineteenth century and up to World War I include the following: a humanistic moral code, competitive individualism, the nation state, reason, the development of science, Progress.[4] The first three of these items have in recent decades diminished in influence; the last three, those most intimately related to technology, have grown in weight. And the last of these, Progress, has acquired something of the power of a world-symbolism in its own right. Contemporary times are characterized by a transition away from the humanistic moral structure which clearly supposed wisdom concerning the specifically human good to be at the top of a hierarchy of values ordering men's activities. A quite different kind of good, a Comtean good, came to be sought. Traditional wisdom lost its eternal objective envisaged by Christianity and its function as guide to man in moving toward the fullness of his being. Wisdom acquired the sense of long range cleverness in manipulating the means to ends. The Cartesian-scientific reason, which modern Western cultures strive to develop is not a reason skilled in the consideration of values and of the ends of life. Consequently, the values and ends of life tend to be vaguely defined by uncriticized customs. In addition, the identity and nature of these ends become more and more obscure owing to the greatly increased size and complexity of societies.

In comparatively small societies, the artisans, warriors, and priests form easily distinguishable classes; confusions and tensions in and among the first two could be resolved by appeal to a superior wisdom presumed to be embodied in the priest. In the Platonic Republic, the functions of producing, defending, and legislating belong to each art, but they remain distinguishable in the rather simple order and interrelation among the arts. Order in the whole structure could still imaginably be produced, maintained, and brought within the grasp of common understanding by the philosopher-king in the light of his insight into the Good. But in modern times the monstrously increased size and complexity of societies has produced a radical, far-reaching, but more or less concealed, alteration in the prevailing vision of the values and ends of human life.

The point is frequently and cogently made that even in a single industry the machinery as well as the organization by which it is operated and administered, have become so complicated that only a few specialists have an adequate grasp of it. Add to this the fact that machinery, which the worker is occupied in producing, is continually replacing the worker. Even the decision making function of the managers and directors is being taken over by machines so complex that only the specialized engineer understands them. Thus, in industrial society more and more complex mechanical or administrative machinery is placed between the operator at all levels and the end toward which he is working. Likewise, a similar development is evident in scientific research work. Important projects are mapped out by (governmental) scientist-administrators and different parts of the project turned over to a hierarchy of subordinate groups. Eventually team work in the laboratory executes some bit of research work related remotely and by a complex and obscure route to the final result. My point is evident; the end envisaged, which all the work is for, grows more and more remote and indistinct. The work in which men actually engage, accordingly, becomes more and more routine and mechanical, less and less the play and utilization of their full powers manifesting their being.

Where the end and purpose of a whole complex process thus becomes ambiguous and remote, what do men actually take as their end? Two answers to this question have been proposed. Some observers believe that nothing at all is effectively taken over as the end and purpose of work. The worker, in consequence, becomes alienated; he suffers a sense of purposelessness, of enslavement to a machine which he cannot understand and to a social order whose values he cannot share. The more thoroughly men are involved in such a seemingly pointless industrial complex, the more frustrated they become and the more ripe for rebellion. Still, despite the persuasive power of arguments of this kind, the revolution predicted does not come off as expected. Thus, other observers have concluded to a second answer to the question. They hold that other psychological mechanisms are effective which adjust the worker to the frustrations of super-industry. An important one of these mechanisms is the habit of converting a means which one frequently has occasion to use into an end in itself. For example, a young man who first goes to sea in order to make a living or to pay off a debt, comes at length to feel comfortable in his nautical surroundings and ends by going to sea as he goes to his home. Life on and with the ocean has become the end and function of his life. He *is* a sailor. Perhaps man's life in the technological world undergoes the same kind of metamorphosis. Men of industry are persuaded to develop the social character by reason of which they come to feel that the rewards of that kind of life overbalance its frustrations. Super-industry, technology, and their continual elaboration or growth, thus, become not a means to human life, but rather its end. Men who live in and with technology, who invent, build, and make it grow, thus acquire the character which finds in this mode of life the fulfillment of their lives. Without it they feel impotent. With it they seem to possess an undeniable value. Their participation in the development of technology becomes a value, a given not-to-be-questioned in which, by a happy fate, they are allowed to participate. Other times, other countries seem to them to be backward, uncivilized, underdeveloped.

My suggestion is that the notion of a technological society begins in the minds of its initiators as a project of continual

progress toward some happy completion which modern
technology has for the first time brought within range of the
practical man's pursuit. Then this technological means undergoes a metamorphosis into the end. The technological
means comes to be seen no longer as an instrument for attaining
a desired form of life. It becomes, rather, life itself:
technological growth, industrial progress, come to be the
justification of life. This end, moreover, may come to be
revered by a whole society. Men can be persuaded to devote
themselves to this cult, even though under it they are no longer
regarded as ends in themselves. No longer are machines for the
sake of man. Rather, technological progress itself is taken as the
end and as the obvious justification of all men's endeavors by a
fully technological culture. Essential to technological culture,
then, is the profound and unquestioning acceptance of this end.

Another route to this same view is worth indicating. Consider
the remarkable growth in independence of once subordinate
techniques of production. These techniques were at one time
elements within a rather simple institutional framework. Their
contemporary practices were legitimized by appeal to the
prevailing myths or world-beliefs which provided the
justification for all human activities. Later these same, but
more developed, techniques became subordinate
organizations—arts, guilds, "mysteries," trades—within institutions whose ultimate justification was again provided by the
prevailing world-beliefs. In modern times, though, productive
techniques, empowered by the application of scientific
knowledge, have become independent both of their parent
institutions and of inherited myths. The techniques of steel
production, for example, work as well for the Japanese as for
the Americans. Justification of the production procedures is
effected by appeal to laboratory findings and to an underlying
ideology of efficiency rather than to the precepts of the Shinto
religious faith or to Christian dogma. These modern criteria
seem today to be altogether beyond question; they superceded
the older justification by appeal to myth and to world-beliefs
just as the new productive organizations and the beliefs which

justify their methods and procedures are tending to expand so as to include the old institutions and their beliefs, or what remains of them. Herein precisely lies the contemporary revolution leading to technological culture: the human interrelations once mediated by common language and beliefs about man's nature and destiny are being replaced by a technological model of society in which the role or function—not the individual—is the basic element (we shall return to this point in §37). Here relations will be "rationalized" on analogy to a self-regulating machine. And here efficiency and technological progress will be the guiding ideals. Thus, by this route, we again reach the view that the final value in a technological culture is not man but technology itself.

This end, moreover, is progressive. For one thing, productive machinery can be made continually more efficient. For another, the pressures from population increase and the accompanying needs provoke a continual drive to improve the efficiency of production and distribution. Again, resources of the planet are not infinite: hence, new fuel and raw materials must continually be discovered in order to satisfy the gargantuan and accelerating appetites of machinery. Thus, in its movement toward increased productivity and efficiency, technology changes the natural environment and, of course, the social and economic environments as well. Then these changes bring about alterations and improvements in technology and so on in a continual objective dialectic. This movement of progress might conceivably be slowed only were it to become evident to politicians and industrialists that the processes and products of industry seriously damage the environment and endanger the survival of themselves, society, and machines. The contrived slowing down of technological development must not uncritically be assumed to mark an important change, for such an effort to control this development is itself a technological move. The control of technology will inevitably tend to be seen within such a culture as another problem for technology.[5]

When society as a whole devotes itself to an expansive technology as the end and justification of its whole existence, then I shall call it a "technism." Clearly technism must adopt powerful means of persuasion, but not necessarily, as August Comte concluded, those associated with contrived religious practice. And it seems to have discovered such means (i.e., effective world-symbols) in some Communist countries. In any event, when technology does achieve this semi-religious status, then the culture producing it has become a technism and has settled for Progress as the indubitable and dominant element in its culture. Technological progress comes thus to be taken not merely as an end but as the ruling end. It becomes not merely an unquestionable element in culture but the keystone of the whole interpretative structure. Here we raise the question where this movement of Progress may end. Will it, perhaps, bring us into another world? Is technism a new world?

Consider that the metamorphosis of means into end is not a minor alteration but represents a basic change in the organization of society and of human character. This factor is a shift in attitude which affects men's habitual evaluations and colors their whole world. This change in attitude renders it possible and desirable to regard all objects as objects of science and of technology. All are held to be elements in a calculable system of energy interchange. Nature as a whole is conceived as a vast mechanical-electronic.nuclear mechanism. Man-made machines represent contrived alterations in the local working of this mechanism which lead to results desired by men. Within this vast natural-social mechanism, technological efficiency rather than man's completeness becomes the end commonly pursued. When these changes in attitude and valuation become complete and pervasive, then I suspect Western man will have entered a new world, a world which will manifest new relations of self to self, to others, and to nature, new perceptions, and new attitudes.

For illustration, note that the prevailing attitudes under technism are those which serve technology. Objectivity is an essential one of these. Commonly today one finds objectivity alternating with other attitudes or even dominated by them. For instance, a physician's respect or sympathy for his patient would usually overrule an inclination to use the patient in a possibly dangerous experiment. But in a culture under technism, all is sacrificed to Progress; the ruling attitude naturally becomes the attitude most favorable to this Progress. This is the objective attitude. This scientific or technological attitude can be extended to all other contexts, even to contexts whose elements are persons. This attitude, moreover, moves still one stage further. It habitually regards nature, the system of calculable interchange of energies, as resource at the disposal of the appropriately trained man. Sciences, then, become the means, the measuring and calculative means, for utilization of natural resources. They tell how to convert raw energies into energies utilizable by industry and how to construct ever more complex and powerful industries. When, therefore, technological advance becomes the ruling interest, then everything is taken as object or is to be objectively regarded, and all objects become fuel for further technological development.[6] An object cannot appear to a person who is genuinely an inhabitant of this world except as a system of objectively viewed energies which are somehow utilizable. The final stage of this movement of concepts is reached when man himself comes to be regarded as a system of energy interchange whose behavior can be made the topic of scientific study with the eventual intent to predict, control, and utilize it. Computer man in the totally controlled society is, thus, the end product of technological culture (cf. §36).

It is essential to note that in Western cultures apologists for technological development have heretofore been accustomed to justify it and its continued absorption of more and more of human life into its structure by reference to humanistic moral standards; modern technology feeds the poor, so it is said, clothes and houses the needy, and generally enables men to lead more humane lives. Under technism, however, the tables are turned. Feeding the poor, generally improving the lot of men, is to be justified by the contribution these humanitarian activities make to the advance of technology. And this advance is felt to be self-justifying. Thus, a kind of scientific-technological morality comes to replace the Western humanistic tradition of autonomous man who, however puzzling, believed he was motivated by carefully considered obligations to his own and to his fellow-man's existence and fate. Man under technism loses sight of the humanistic source of obligation but discovers a new source and justification of motives in a positivist and technological rationality.

In sum, technology is productive activity which is controlled by consciousness and reason. It becomes modern when the reason involved is subjectivity disciplined to measurement and calculation with an eye for efficiency. It becomes a technism when efficient productivity is genuinely and deeply felt to be the end of man and society. In comparison with this end, individual man is nothing; his salvation, his completeness, consist in his submission to the demands of a progressive technism. At just this point, where not man but Progress is regarded as the intrinsic value, where the self and its changes are subordinated to the demands of technological evolution, and where nature (including man) is envisaged as a system of utilizable energy interchanges, at just this point man of the Western world faces radical change. We have stipulated that a world is essentially characterized by the view it takes of man and of his possibilities. Here this very basic view is undergoing change, and its inhabitants could discover themselves threatened with a shift from the Western world into another. Here the world-symbols, which hitherto have guided the interpretation of human life, are in danger of being completely lost.

How should so radical a change be regarded? I suggest it may be looked upon in two more or less complementary ways. Either it may be regarded as a nihilism a destruction of the bonds joining self, others, and nature into a whole where all may manifest their being, or it may be seen as a movement into a quite different world, one envisaging a very different set of possibilities and cultivating a quite different sort of bond among self, nature, and others, and where, therefore, another kind of being is manifested. In this world the center of concern will no longer be man's existence and fate; this center will have become the elaboration of technology. Man will have become accessory to manifesting the power of the technological machine. Upon either alternative, the humanistic evaluation of technism is quite clear: technism neglects in principle man's primary obligation and substitutes another; hence, with the advent of technism, the end of the Western world is announdced.

With the new envisagement of man's primary goal and of the obligation characteristic of technism, the other distinctive values associated with the Western world will change radically. Notice has just been taken of one consequent change: that humanistic morality will be replaced by a technological morality. Individualism likewise will be rejected in favor of the corporatism favorable to the management of complex enterprises. In addition, humanistic reason will disappear and its function will be taken over by computers, computer programmers, and their directors. And the necessity for developing the sciences and the scientific use of reason will determine the nature of educational institutions. Scientists or would-be scientists will seek administrative posts in educational institutions and will exert themselves to reform curricula. Secondary schools will then become indoctrination centers; universities will become knowledge factories and training schools in the service of technology and industry.

The consequence of the decision to place human development in the hands of technological specialists will, I suggest, be irrevocable, for when the genetic, chemical, and psychological means of forming and educating human beings are delivered over to a well equipped and objectively thinking managerial bureaucracy, the world actually produced can be made, by application of available techniques, to seem to be the only desirable and right one. In effect specialists and bureaucrats will have been given the tools for manipulating the culture according to their wishes. Possible alternatives, if they occur at all, would be immediately branded as barbarian, revisionary, anti-scientific, mediaeval, or existentialist, and accordingly eliminated.[7] There is no turning back along the road of such Progress. One aspect of the suggested picture of a fully technological culture is the implication that, in effect, human freedom would be neutralized by technological means. Consequently, the assumption that man is non-free, made by those who hold to doctrines like these, could issue as the effective conclusion from this social experimentation. Technological society would then produce the kind of man of which its theory of man is true.

Perhaps not many observers believe our present technological culture has as yet approached this pass. Our culture, nevertheless, presents points in common with technism, so that a passage to technism is indeed thinkable. Thus, our advanced technological culture may be called a limit culture. It borders on another world.

I underline my evaluation by repeating that modern technological culture could lead easily into technism. Let it be added that such a regime is totalitarian. Totalitarianism does not operate to bring all social and individual forces under the control of some privileged person or institution solely by the instrument of terror. It works even more effectively to this end by application of scientifically designed means of persuasion and conditioning, for these means can lead men to desire this form of life. Under so-called benevolent totalitarianism, the enslaved population cooperates in its own enslavement and points with pride to the higher standard of living it has acquired in exchange merely for its freedom, dignity, and the right to pursue the non-objective dialectic in obedience to man's primary obligation. [8]

## §36. *The Technological View of Man*

Up to this point we have pictured technological man merely as a sympathetic and cooperating member of technological society. And here a question occurs: what kind of man would agree to inhabit this new world of technism? Who would willingly be programmed—so to speak—to being a minor machine within such a machine? Now, technological man himself has a theory about man which, perhaps, answers this question. Remembering that a culture reduces to its own terms everything encountered within it, we shall expect the vision of man belonging to this culture to be of a piece with its abstract nature. This section will be devoted to describing such a man and to evaluating him with respect to the primary obligation courageously to pursue the truth about oneself and to seek to understand the meaning of the essential human events.

Technological man belongs to a lineage which begins with the spiritual automaton of the Renaissance and reaches toward the electronic automaton or electronic-human automaton of the predicted future. To understand him as he is presently developing we shall have to ignore the peculiar sense in which a man is said to be a unity, the irreplaceability of the self, and many other of the properties traditionally attributed to persons. Furthermore, we must agree to limit acceptable information about the self to that obtainable by numerical measurements and by the operations upon measurements of the type allowed by physics and chemistry. What now will be the account of oneself or of others? Just such an account of man has been extensively elaborated in the mechanistic tradition from Democritus to B. F. Skinner and is receiving its most interesting theoretical development today, not in psychological, socio-economic, or biological theories of man, but in what I shall call the computer doctrine. A brief glance at this specialist doctrine will be interesting to us for several reasons; it will provide a contrast with the account of man as he commonly experiences himself (discussed in Chapters II to V); it will provide an indication of the way in which many scientist-technicians are inclining to feel about themselves and to interpret human being; likewise, it will serve as just suggested, further to elaborate our criticism of the technological culture which tends to accept a doctrine of this general type.

To an astounding extent the operations of the human mind, at least those of the discursive mind, seem to be imitated by modern calculating machines. For a long time these machines have been able to perform complex mathematical operations with lightning-like rapidity. More recently a computer has been

designed to play checkers. It looks a number of moves ahead
(i.e., imagines); it distinguishes between good and bad moves
(i.e., evaluates); it remembers and reuses its good moves and
good types of moves (i.e., learns from experience and
generalizes); it works harder to beat a good opponent than a
poor one (i.e., exhibits the competitive spirit). This computer
now consistently wins against champion players. Moreover, a
method has been developed by which this machine can teach
other machines to play the game. Another machine can prove
theorems in symbolic logic using something like the general
methods and heuristic rules which humans use. Still another
can compose music which some listeners find aesthetically
pleasing.[9] Also a machine has been programmed to diagnose
symptoms in a manner which agrees with consulting physicians.
A self moving machine has been constructed which will obey
simple orders (e.g., to go to the third room on left and get a box
off a platform; it must figure out a way to get on the platform
and to manipulate the box).[10] These and other astounding feats
seem to cover the range of human intelligent action. I need not
further extoll the virtues of these machines but will merely note
that since they seem to produce the same results as man, they
are often said to be his "operational equivalent"; they exhibit
his technological measure. Let us consider a little more
carefully just what it is which such machines measure.

The question is debated whether (or perhaps when) one
should refer to these machines as human; it is most seriously
posed by those philosophers in the Cartesian-scientific tradition
who identify the essence of humaness with the kind of intelligence best exhibited in mathematics. Men of this sort have
thoroughly disciplined their subjectivities to the demands of
quantified reason. Their apologist philosophers sometimes
appear to argue in a rather tight circle. They begin by defining
man as the intelligent animal, then in effect they define intelligence in terms of the step by step operations which can be
built into a machine. Thereupon they construct a machine
which performs these operations and then draw the inference

that the machine is virtually human. Their more serious arguments, however, begin from a definition of artificial or "computer simulated" intelligence. [11]

Computer simulated intelligence has, for this purpose, been defined as behavior which, if exhibited by a human being would ordinarily be called intelligent. Some of the accomplishments of the computer suggest a close similarity between artificial and human intelligence. At least the results, at first glance, seem to be the same. Humanistically inclined philosophers, however, often point out that terms like information, language, imagine, remember, learn from experience, generalize, prove, and the like are employed only metaphorically when referred to artificial intelligence and are on a level with personifications not infrequently used jokingly in the physics laboratory. But their opponents are more serious; they ask the humanistic philosopher to point out the exact aspect of human intelligent action which requires a different use of the terms just mentioned. Whenever such a philosopher seems to succeed in describing unambiguously an action of this kind, the computer technician proceeds to devise a new program or a new machine which can perform the given act. Baffled, the hunanistic philosopher may take refuge in asserting that men are able to do quite different sorts of things, things which a computer admittedly cannot do; for instance, a man can enter into friendship with others, be charitable, decide to attempt the impossible, become angry, and engage in like human acts. Sometimes characteristics such as these are said to belong to the *way* a man is or exists rather than to the operations he performs. Thus, it is said that a man's being is different from a computer's. In particular, even though a machine can add two and two and get four, it does not "see" that it does so (cf. §18), much less does it do so with satisfaction, in a friendly manner, or even with indifference. It does so mechanically, for it has no subjective being. But, as we saw in Chapter IV, it is extraordinarily difficult to say non-metaphorically what one means by the experience of subjective being or the experience of (say) seeing. Such seeing cannot be quantified and measured.

Rather, it is presupposed by the technologist's operations of quantifying and measuring.

To this point the apologist for the technologist has devised a rejoinder. We have no way of access, he will argue, to any object except by way of observing its behavior; if the way a man is or exists is not exhibited in his public behavior, then we have no mode of access to his being or to the way he exists and can make no verifiable assertions about it. Hence, we need give our attention only to his behavior. Now, one essential mode of human behavior is mental calculative behavior, and this behavior is duplicated by the computer.

More specifically, the argument is put this way: so far as human behavior can be observed and accurately or clearly described, the description can be formalized and put on tape or punched on cards; then a computer can be made to simulate it. But if human behavior cannot be clearly described and formalized, then we do not clearly know what it is. Hence, either human intelligent behavior can be simulated by a computer, or we must be altogether vague about it. Evidently the adherents of the computer view of man answer in the negative the old question whether a man is something—e.g., a substance, an ego, a subject, a self—in addition to the collection of operations he can perform. Their view may be expressed in terms of operational equivalence: if X can perform the same operations as Y, then X is the operational equivalent of Y. Now a computer can perform the supposed unique and characteristic intelligent operations of man; hence, it is his operational equivalent.

There are two instances of loose verbiage in this argument. The usage of "operational equivalence" is one of them. Operational equivalence does not mean identity. Even if a computer and a man produce the same results, it does not follow that they are produced by the same operation. It is not efficient to use the phrase "operational equivalence" so loosely that one cannot distinguish between two computers using the same circuits, between two men playing the same role, or between a computer and a man. We can and do make these distinctions.

The difficulty concealed in the phrase "same results" is more subtle. What counts as the result of a man's computation when he adds (say) two and two? Just the response, "four"? But in fact a man never delivers exclusively such a response. His response is always communicated "in a definite way"; e.g., with boredom, with excitement, with hostility, with objectivity, or the like. Now, it makes no sense to say of a computer that it gives or does not give its response in any such *way*. Its "answer" is simply that part of the electronic process which especially interested the operator. And feelings form no part of this process. Or do they? Today this question is seriously debated.[12] The most sensible view of the problem, in my opinion, is the following. The human activities which are susceptible of being formalized and programmed, and thus of being imitated by a machine, are processes which move by a discrete series of steps to a final completion. Or else they move by approximations to determine the values of continuously changing variables. Moreover, such processes reach their completion stage with more or less success. Such are the behavior processes involved in solving a problem, playing a game, engaging in a productive technique, and the like. Human powers, however, are not exhausted by these processes. Some capacities do not involve this step by step procedure. Feelings, hunches, moods, emotions, pains, belong to this class of capacities; none of them comes into being or reaches completion by a step by step procedure or process; none is accomplished with more or less success. Each is simply present or not present. Each, like the sensation of red, is either had or not had. Now these experiences (feelings, etc.) which are immediately present cannot be programmed.[13] There is, in such experiences, literally no behavior, no process, to program. If a computer were somehow to be made to experience feelings, moods, and the like, it would do so not by means of a novel and clever program, but rather in consequence of some radically new construction of a type not even remotely envisaged at the present time. Of such a "machine" it would probably be more reasonable to say that it was human, than to say of a man that he was such a computer or a machine.[14]

Today, in order to compare a computer's response with a man's, the man's "way" of responding must be abstracted and ignored. Thus, all that which we earlier included under the heading of felt possibilities must be left aside. And since a self (in the sense of this essay) is developed by interpreting felt possibilities (the non-objective dialectic), a computer cannot be said to have or to develop a self. It imitates only man's problem-solving or task-achieving activities. It should no more be regarded as a man than should a ditch-witch, which admirably imitates a man's ditch-digging technique.

Experimenters who claim to have achieved computer simulation of human feeling, mood, emotion, etc., in fact abstract from these experiences. But their abstraction is camouflaged. Thus, a machine programmed to simulate a mental illness, "conversed" with a psychiatrist and fooled him into thinking the machine was a human patient. The machine and the psychiatrist, however, communicated by teletype. By means of this camouflaging device, the psychiatrist's characteristically human way of responding was filtered out and the machine's failure to communicate in a similar way was concealed. Thus, the communications were carefully limited to provide that which the experimenter wanted to discover. But this circularity is too patent! If we know ahead of time what kind of results the machine will yield, and if we limit the man's response to just this kind of mechanical result, of course they will seem to be comparable. But in fact, a comparison between the man's response and the machine's was not made at all. Rather, the results of two teletype machine operations were compared. Here the machine measures the man only after the human character of the man's response has been removed.

Paralleling the distinction between the programmable and the non-programmable, there is a distinction in language use, (cf. §33). Consider the question: Are the precise (or literal) languages — i.e., those which may be formalized and placed on computer tape — adequate for the description of all human being, activities, or operations? A negative answer to this question may be approached by observing that we do not have the slightest idea what language the human brain, considered as a computer, uses; [15] hence, we cannot judge in advance concerning all its possibilities nor whether they may in the future all be duplicated by a machine. Besides, not all formalizable questions about the total possibilities of a complex language can be answered without paradox, as the Gödel theorems indicate. A telling point against the conviction that all human responses can be formalized, however, is this: many of the most typical human experiences, such as experiences of introjection, decision, anxiety, insight, self-awareness, self-identity seem in fact to be resistant to precise and literal description even in languages having the richest logical resources. The immediacy characteristic of these experiences (or aspects of experience), as we noted above, offers nothing to formalize. Thus, the fact is that the formal languages are insufficiently powerful to describe these experiences. Hence analogy, metaphor, suggestion, ambiguity, irony, myth and other means for enlarging the powers of literal statement are constantly used by any one attempting to do justice to these characteristically human phenomena. Conversely, languages which permit enrichment by means of such devices at the same time allow expressions which are not formalizable—at least not in any sense now current—and which, therefore, cannot be put on tape and communicated to anything now called a computer.

The sense of being bound by ideals, by acceptance of values, or by feelings of obligation or guilt are typical experiences or aspects of experience. They are immediate and are not formalizable in patterns of discrete units which can be punched on cards. Awareness of these feelings can be described only by the poetic and dramatic means and communicated by evoking something of the feeling denoted. A computer, which can act as programmed only upon reception of discrete bits of information, cannot be instructed to act upon these immediacies; hence, it cannot sense the obligation to pursue the truth about itself or to discover the meaning or value of the essential human events. In this respect, the computer doctrine of man offers an inadequate account of the whole self. The machine in principle cannot take the measure of man, and technism, which would cut man to this measure is defective in its vision and does him violence. Thus, I return to the conclusion: our negative evaluation of the technological culture or world was in this respect not exaggerated.

This exposition will probably have brought a question to the reader's mind: if the computer cannot simulate certain aspects of the human use of language, why is this inability not in fact noticed by everyone? I propose now to respond to this question, and then to make a general remark about the contrast between the humanist and the computer measures of man.

More exactly, the point to which I need to respond is the following. Every one agrees that language can be used objectively. When so used, the feelings and other personal qualities of the users are not relevant to the meanings communicated. But if the unprogrammable ways in which all human communications are expressed are an intrinsic part of these communications (as I have asserted), then by what means can they be abstracted or omitted in language used objectively?

My answer to this question is that the unprogrammable how-a-communication-is-to-be-understood can never in fact be omitted from any human occasion of language use. Nevertheless, the attitude, the feeling involved, the *how* of speaking or of understanding, can be standardized. Such a standard attitude is always relative to a context. When the standardized attitude is utilized, members of the appropriate language community who share in the given context immediately recognize the attitude and respond automatically. The exclamation mark and the question mark refer to and evoke such standardized attitudes. Some such attitudes, moreover, do not require any overt indications. Their presence or the need for them is immediately recognized. For example, in the scientific community the attitudes and feelings of objectivity and impersonality have become standard. Scientific reports and remarks expressed by those engaged in the cult of techniques are announced and received, not without feelings, emotions, and attitudes, but only with those which have become standard and thus pass unnoticed. They are not noticed, just as the customary local speech accent passes unobserved by members of the local speech community. Or again, they are not noticed just as many of the properties of lived space or time are ignored (not annihilated) when they are being quantitatively measured. This standardization evens out the differences in emotional and attitudinal response among men, rather as the differences among concrete individual persons are evened out and ignored when a group of persons are counted and assigned a number.

The machine measure of man measures only his mechanical characteristics. The *how* of behavior, speech, thought, being non-mechanical, must be standardized and systematically ignored when this measure is used. The image of man which emerges from this abstractive process is a far cry from the human self, involved in concrete life, a far cry from the man engaged courageously in discovering the truth about himself by means of mimetic action and the non-objective dialectic along ways designated by the world-symbols of the West. The computer image, and kindred images, of man express not so much the truth about the whole self as the desire of some individuals to predict and to control the behavior of others.

This discussion is not intended to suggest that the issue between the adherents of the computer view of man and those who hold to a more humanistic one is or ever will be settled. It is difficult even to grasp the issue between these two. The setting of this supposed disagreement is rather too reminiscent of the battle between the elephant and the whale: a battle which never took place since no common context could be discovered wherein battle might be joined. In order that there be an issue in common, there must at the minimum be a language in common. But for one of the contestants the linguistic ideal is formalizable, literal expression, and for the other the ideal is insight awakened by ambiguity, metaphor, analogy, and myth. Here there is scarcely a common language. Thus, the situation is not so much that a clear difference between the behaviorist and the humanist has not yet been formulated as that a precise issue between them is not formulatable in a technological context. This difference was pointed up in our discussions by the scientist's appealing to overt behavior and special experience for evidence and the humanist's appeal for evidence to our common experience of human being, to self-awareness, self-questioning, birth, death, and their meanings. And there is a correlative difference in the two kinds of practice. The humanist can provide only hints of an invisible maturing of the self as the outcome of his non-objective dialectic; whereas the man under technism can point to the very palpable and inescapable alteration of the whole environment and of our conception of man's place within it as the consequence of the practice of the objective dialectic.

The scientist is suspicious of the humanist's data because they cannot be precisely measured and translated into the clear, numerical terms which he finds intelligible. But upon reflection this suspicion appears to be unreasonable. It is rather like expecting a measurement made with an ordinal scale to have cardinal properties. The scientist is interested in man as a unit similar in some measurable respect to other men. The humanist philosopher is interested in him as a unique being, a "one" in the fifth sense of number. It is scarcely extraordinary that the account offered by either of these is unsatisfactory to the other. Specifically, the humanist philosopher must reject the measurement of man by the computer doctrine, not so much because it is false as because its validity is confined within narrow limits and is irrelevant to the sense of "man" with which he is concerned.

Perhaps the best that can be accomplished at the moment toward settling this apparent issue is to place the two views sharply side by side. This mode of "argument" was adopted by Plato in the *Gorgias;* there he placed verbal portraits of the philosophic and the despotic life in juxtaposition and, as it were, asked the reader to make his choice. In effect, the judge in this conflict is common experience or common feeling for the possibilities of human completeness. And the verdict of this judgment, in the present case, is clear: the computer or technological measure of man is not valid beyond the laboratory situation for which it was devised. With this conclusion we put aside the measurement of man by the computer doctrine. Instead, we turn to consider the conditions under which technological man envisages his role.

## §37. *On Playing the Technological Role*

Even though a man may not be consciously and critically aware of the primary obligation to discover the truth about himself, still the conditions for pursuing this goal may be present. Are they present in a technological culture? How does technological man, who regards himself primarily or only as a means to the advancement of his culture, and as an object to be studied and manipulated like a computer, experience himself in relation to these (dramatic) conditions?

Typically the scientist-technologist is a practitioner of the scientific method within various contexts. His role, then, must be understood in relation to this method. But the *way* in which he plays this role refers us to his immediate experience of himself in discharging his function and requires of us an interpretation of his self-identification, of the way in which he carries out his decision, and of the insight which he finally reaches. We are interested in the way in which he lives his role and need to ascertain whether he conceives of his task within the limits imposed by the human condition and is influenced by Western world-symbols.

It has become a part of folk wisdom today to know that a scientist, recognizing a problem, formulates a hypothesis which might answer or resolve the problem. He then draws empirically verifiable inferences from that hypothesis, perhaps with the aid of complex measurements and mathematical techniques. Next he constructs instruments, if necessary, and sets up a controlled experimental situation, wherein the hypothesis may be tested. Finally, he compares his inferences with the actual results obtained from his instruments or the controlled situation and concludes to the acceptableness or erroneous character of his initial hypothesis in this or that particular. This process, the objective dialectic, is represented as a kind of interrogation of nature or it is said to be something like measuring the hypothesis by nature. Throughout the testing phase the scientist, of course, maintains a stardard impersonal and objective attitude.

Obviously, in this brief sketch I have vastly oversimplified the practice of the scientific method. Phrases like "set up an experimental situation" or "test an hypothesis" may refer to years of work, perhaps on the part of many workers. Nevertheless, an oversimplified account of this method is useful in that it emphasizes certain pervasive factors. For instance, it exhibits the general pattern of formulation of hypothesis, experimental trial, and confirmation or disconfirmation as clearly analogous to the role structure of decision, struggle, and insight. Also, it underlines the scientist's function as the objective and impersonal observer.

To say that nature herself through mediation of the scientist's objective mind, tries or tests the scientist's hypothesis is evidently a metaphor. This manner of speaking is characteristic of the earlier history of science before many poetic expressions and views had been purged. "Nature" in fact is the environment as it has come to be understood by the scientist. Thus, it is not an uninterpreted nature herself who tests the hypothesis, but rather a man, a student, whose subjectivity has already been formed by a highly specialized discipline, who is disposed to certain attitudes, and shares the prejudices of his culture, who tests the hypothesis and draws the soundest conclusions his criteria and skill allow. Now this man, the scientist, in practicing the objective dealectic is relating himself to the whole pattern of hypothesis formulation, experimental testing, and confirmation or disconfirmation. It is he who passes judgment upon the outcome, he has the insight, so to speak, for the natural objects or events which he studies, and decides whether the initial decision or hypothesis is appropriate or needs reformulation. The crucial point should now be clear: under technism it is not concrete man whose career and whose change is the central fact: rather, this center is occupied by an abstract nature or by a scientifically developed concept of nature. Not man but his present environment is to be scientifically understood and changed. Not individual man but an abstract vision of nature is the hero of the drama. This environment consists in the abstract space, time, object, and the man which scientific and technological investigators have, over recent centuries, gradually forged[16] and are now utilizing with ever increasing efficiency and devotion. Our task, therefore, is to understand the way in which technological man plays his roles under these notable alterations in the dramatic conditions.

My concern here does not touch the psychology of the scientist but only the conditions of role playing. Our discussion of role concluded (in §§15 and 23) that the role-player, like the hero of ancient tragedy, was possessed of hybris, an ontological defect, and made, therefore, a faulty decision about himself, struggled to enact the decision, but being tried by fate, reached an insight into the error in his self-identification. Here we need to ask how the scientist-technician experiences himself in this pattern. Our task will appropriately be divided into three parts to correspond with (A) role choice, (B) trial, and (C) insight. My analysis of the role of scientist in a thoroughly technological culture has led me to the conclusion that in the way he conceives his role he fails to recognize the dependence, non-identity, and finiteness of the human condition. Reasons for this conclusion will now be offered.

## A) WHO IS THE MAN OF TECHNISM?

We have noted that the scientist's hypothesis undergoes trial by the scientist in his laboratory, but the scientist himself as a person is not put to the test. Also, the scientist's cultural view embodied in the scientist himself and within which his role is identified, is not ordinarily the subject of decision and testing, even though the particular theory within which his hypothesis is formulated may be tested along with the hypothesis. The scientist decides to stand aside in his own person from the pattern of radical change and to judge of the identities and laws assigned to objects. Furthermore, as technician, he plays the role of disposer and designer of nature, and, at the same time he is the servant of the technology which controls nature. Thus, technological man's decision is to play a role which on its theoretical side requires that he be the judge of nature and on its practical side requires that he be the disposer of nature and the servant of his machines. This complex decision is of great moment. Strictly technological man pursues the truth not about himself but about nature and its possible uses. He chooses a role which in some respects is very like the role played by fate in ancient times; in other respects it is rather like to the role of a servant, of a human instrument, even of a victim.

A number of philosophers have concerned themselves with the alienating effect of technological man's role as the servant or as the victim of machines.[17] There is no need to repeat their observations here. Instead I shall devote a few remarks to that aspect of his role in virtue of which it resembles fate. Hence, I emphasize the analogy: the scientist-technologist, or objective observer, is to the experimental method as fate is to the ancient drama.

This analogy recalls an earlier turn in the history of understanding fate. In Ancient Greece, fate which stands in judgment over men was first envisaged as the dictate of an obscure but an independent and impersonal power (Dike) which determined even the gods; later this power was psychologized, so to speak, and was understood as the will of the gods, especially of Zeus, perhaps just because Zeus held the power of prediction and control. In either circumstance, fate was independent of the world and of ordinary mortals in it. Early in modern times the objective element in experience which seemed to pass judgment upon hypotheses, was visualized as an impersonal and mechanical nature or as the laws of nature. Again, this element came to be psychologized and was envisaged as the inevitable and objective judgment of the scientist. Nature, that is, was experienced and thought as subjected to the scientist's theories. The scientist's aesthetic sensibility, even, came to be recognized as an important item in the determination of such theories. By this shift, the environment seems to be brought far more intimately within the power of the scientist: his invention became the mother of necessity; the scientist-technologist, possessing the power of prediction and control, can remake the environment and subject it to his will. Fate, thus, seemed to become a function of man; it is that which the scientist's measuring schemes and their theoretical interpretations can translate into a publically intelligible language; it is that which the technician decides to bring into being upon a plan of his own. Man thus becomes the maker of nature-as-measurable. At the same time the experimenter's sense of himself being in a process of change is diminished. Perhaps other people are the topic of his investigation or manipulation, but he takes an objective attitude toward them. He supposes himself, the experimenter, who measures and passes judgment, to be an independent and external observer, untouched by the processes he is measuring.

There are humanists who are ready to maintain that the role of fate, or of judge and disposer of nature, is an inappropriate one for a man to attempt, whether he be scientist, technologist, or whatever. Indeed, they hold such an effort to embody an exaggerated claim to independence and to be an exhibition of hybris which the possession of a mathematically disciplined mind does nothing to ameliorate. Still, it may seem unfair hastily to condemn the scientist's decision to identify himself as the fate which tries and judges hypotheses about the objects of his science.[18] One might equally well condemn the physician, of whom everyone approves, for playing a fate-like role in relation to his patients as he manipulates the body's functioning with a view to bringing about the desired condition. Indeed, any artist seems to dominate or to attempt to dominate and to determine the fate of the matter in which he deals. This domination is inevitable, although perhaps it betrays the incipient hybris associated with any art. One would begin to object to the scientist's decision to engage in this fate-like role only when it seems to be held without due regard for human limitations and dependence. For the scientist-technologist is dependent, as anyone is, upon the already existent culture into which he was born. And he is limited by the always partial manifestation of his own being which that culture allows. So soon as this dependence is ignored, then the incipient hybris begins to take actual effect, then he identifies himself wholly with his role. He neglects the non-objective dialectic and loses the recognition that his own dependence as a person must prevent acquiescence in any role which assigns an independent, permanent, and fate-like function to the self.

The changes which the scientist-technologist can permit himself remain within the type which accords with his decision. He seeks, therefore, to become more skilled in his specialty, more objective in observing and thinking about nature, more clever in designing machines, and more efficient in using them. He avoids any movement of the non-objective dialectic which suggests the insufficiency or incompleteness of his chosen role. Thus, he holds himself aloof from the pattern of significant self-change. And so doing, he is, as it were, laying claim to being independent of the human condition. Our measurement must, therefore, award him a degree of guilt.

## B) SELF-IDENTITY UNDER TECHNISM

Technological man is the specialist *par excellence;* he has thoroughly identified himself with his role and feels it his highest privilege to contribute to the advance of Progress.

One might observe nothing exceptional or strange in this identification. Geniuses and others too of all times have given themselves with single-minded intensity to a single project. Society has benefited incalculably from their concentration. One must also observe, however, that the compulsive devotion urged, even required, by the difficult and arduous task of developing a technological culture is devotion to the objective dialectic. The non-objective dealectic is minimized in principle. For the objective of this culture is not the manifestation of human powers but the utilization of the forces of nature. Nature, not the human self, is the hero of this drama.

True: one can engage in a productive role or art while at the same time making this role or art a means to achieving insight into the self and so to rendering one's own powers explicit. This feat is just that accomplished by the hero of a Greek tragedy. Quite another mode of doing the same is recounted in *Zen in the Art of Archery.*[19] Still, whatever the success of individual scientists or technological men in combining the specialized and the common needs of man, culture under technism attempts no such unification of personal and specialist demands. Justification for a single-minded self-consecration to technological tasks is, rather, shifted solely to the grounds of social benefit and contribution to Progress. This latter good is deemed to be sufficient. Nor is the famed humility of the scientist usually judged in this culture to be a mode of self knowledge. Rather, it is interpreted as humility in the face of the magnitude of the task of harnessing nature to satisfy the needs and desires of men and to the endlessness of the march of Progress.

In identifying himself with his role, technological man adopts the attitudes demanded by his role. The dominating attitude is objectivity, the attitude of one who seems to stand aside from the flow of circumstances and to judge this flow impersonally by use of the concepts and ideals which underlie the sciences and technology. This fate-like and objective disposition is regarded as supremely valuable and as limitlessly applicable.

The consequence of judging the objective attitude to be applicable without limit is that technological man must direct this attitude upon himself. That is, he must take himself as an object. Objects, however, do not experience time as the dramatic hero experiences it. Objects endure in objective time and do not change, unless in the sense of undergoing alterations in the distribution of the energies of which they are composed. Certainly, they do not change in the sense in which a self may experience dramatic and radical change. Since technological man regards man as an object, the identity which he must, in his own view, attribute to himself is permanent and contradicts the non-identity or openness to radical change which our former investigation required of any self (§12).

Finally, the man identified with his role and wholly given over to his work must fail to utilize and to express aspects of himself not demanded by his work; often the results are disastrous. Usually apologists for technological culture emphasize the leisure provided by technology for "self-development." But this kind of leisure is only a vacant place in the program of work. On the average, non-work time is filled by distractions ranging among sex, soap-operas, civic and other activities designed primarily to renew ability to labor and to reinvigorate the attitudes necessary to technology. When this leisure is not merely a distraction from distraction by distraction, it is bent to the service of technism, not of man. On the other hand, the man who recalls the poetic and dramatic dimension of self (§§10f), a dimension incommensurable with everyday roles, can engage in quite another sort of leisure. Perhaps he will recall that primitive man returned from labor in the fields or from the hunt to engage in ceremonial festivities. For example, African Bushmen begin a ritual dance as soon as the successful hunter appears with game over his shoulder. By means of the ritualistic and mythical elements of these ceremonies, men lose their every-day identities and come into touch with the world which gives meaning to their labor and existence. Leisure of this kind is a sojourn near the gods, a journey out of the Cave, a release from daily roles and preoccupations. By this means the unfamiliar and poetic part of the psyche may be invoked and experienced and the self exposed to the possibility of radical change. It will be remembered that the theater began as a playful mimicking of the gods. Without this mimetic communication with the gods or with the being encompassing the work-a-day world, a part of self remains infantile (§20). Western man, though, under pressure from the demands of technological culture, seems to have lost the capacity to alternate or blend work and festival, purposive labor and ceremony, calculative thought and myth, the objective and the non-objective dialectic.

Thus again, our criterion, derived from the conditions of

selfhood, leads to a negative judgment of the typical man of technism, a man unquestioningly identified with his role. Technological man knows too well who he is.

## C) THE INFINITUDE OF TECHNOLOGICAL MAN

It is important, finally, to reflect upon technological man's relation to the finiteness which our consideration of the meaning of death led us to ascribe to the human condition (§26). Now I believe it easy to see that this man denies human finiteness. This denial becomes evident upon observing that technological man identifies himself with a role which repudiates or minimizes the meaning of death.

To technological man, that which undergoes significant change is his understanding and use of nature. We have observed that nature, the environment, once thought of as the enduring stage of dramatic activity, has become the hero of dramatic change. And, in fact, according to contemporary opinion, the notable changes which modern times have wrought are changes not so much in man but in understanding and use of the environment. Thus, modern man rejected the identification of himself as the subject of dramatic insight; rather, he stepped off-stage and busied himself with study and the redesigning of the props. Having identified himself wholly with the fate-like role of objective judge and manipulator of nature, he himself stands aside from the pattern of change. The consequence of this decision is to alter the way in which his limits are envisaged. As identified with his role, he seems to appear without temporal end, to progress indefinitely into the future. Or to put the matter another way, the death or survival of the individual comes to appear negligible in comparison with the continued execution of the role. The invariance of role, not the variant who plays it, becomes the critical factor.

Still more important for understanding technological man's denial of finiteness is his belief about and attitude toward death. Even if he does not go so far as to believe man to be an ambulatory computer but is content with holding him to be the inevitable consequence of eons of chance variations and selection by the environment, then death in the humanistic sense does not even exist for him. For death from a naturalistic standpoint is nothing more than an alteration of the physical organism, a change from a state of greater to a state of lesser complexity, an arrest of metabolism. Death under technism is no unique event. In general, life, birth, and death occupy no privileged position in a fully technological culture; they are not manifestations of a unique human being. Rather, they are events on a par with all others and may or may not offer timely and interesting scientific or technological problems. Death, then, in this culture would present a problem in respect to its determining factors and their control, but its meaning for the individual who faces it is not a timely topic.

Technological man's cue for understanding the significance of death, therefore, does not lie in considering the way it manifests human being or dramatic insight into oneself. Indeed, the good technophile ought to be too busy with his job to devote thought to so lugubrious a matter as death. But to avoid thought about this strange event is to fail to recognize the nature of human finiteness; it is to fail to grasp the impact of the most basic and fateful threat to human existence. Just the impact of this threat arouses the questioning attitude toward human existence which may, if anything will, discover its meaning.

The man of this culture or world, thus passes over the most effective occasion for evoking the non-objective dialectic. He must admit, of course, that he can become incapacitated or dead. But he judges his individual death objectively to be an unimportant incident in the history of Progress. He feels that his own finiteness is complemented by the limitless power and expanding life of the species engaged upon this saving and Progressive way, as the Christian once felt his impotence to be complemented by the Grace of God. In addition, technology is confidently expected eventually to eliminate death. Actually much medical research now envisages something like this outcome. The faithful as usual look forward to longer, even indefinitely long, life. This expectation, I suggest, strengthens technological man's denial of the finiteness indigeneous to the human condition.

We conclude this part of the chapter by recalling that our examination of experience of the self in chapters II to V led not to a definition of self but to a characterization of its direction of development together with a grasp, at least in negative terms, of the conditions limiting that movement. The self, we concluded, is obligated to itself courageously to pursue insight into its possibilities, within the limiting conditions provided by constant recognition of its dependence, its finiteness, its non-identity with any role and as guided by the world-symbols of the West. But now this chapter concludes that man under technism is the servant of his technology and, therefore, systematically ignores the primary obligation of self-inquiry. Furthermore, we have considered his relationship to the limiting conditions of humanity and have found this man and his culture wanting on each score. Rather than seeing himself as dependent, without permanent identity, and finite, man under technism experiences himself as the fate-like judge and designer of nature, who faces all problems with persistent objectivity, supremely confident that if he himself does not solve them, they will nevertheless yield to the march of that saving Progress with which he feels he ought to be at one. Thus, he systematically avoids the insight which would reveal the deficiencies in his decisions and the possibilities of his being to which he otherwise might gain access.

## §38. *Conclusion: Technism and its Consequence*

Now that technism has been measured by the standards of this humanistic philosophy and the outcome of the evaluation demonstrated to be negative, the appropriate question to entertain concerns the meaning of this measurement. Perhaps it will be suggested that this meaning is negligible, for this extreme form of technological culture would never be pursued by sensible persons. The point is, of course, that this extreme and ideal form of culture does indicate the direction in which we ourselves may be insidiuously moving, and the measurement of it evaluates various aspects of our possible future character. Being thus forwarned we may be forearmed; hence, it is important to ask what our measurement reveals in respect to technism. I want to show that technism is inseparable from a special form of violence. Further, I want to suggest that in a strictly technological culture there is nothing to prevent this violence from becoming destructive of that culture itself.

Not that any culture whatsoever is innocent, but no one will deny that the development of technological culture has been accompanied by an unusual degree and amount of violence. I cite a few of the ways in which this violence has actually become evident and will then indicate its quite special character.

On the large scale social plane the self-destructive nature of the scientific-technological culture has been manifested in matters like ecological imbalance, the population explosion, the energy crisis, war, and the like. And in fact we develop and make use of chemicals which kill insects harmful to crops; in reality we have damaged, perhaps irreparably, natural ecological systems and have set up conditions for breeding insects immune to these controlling poisons. We develop sources of energy whose waste products endanger life and the sources of food both on land and in the seas. We devise methods for preserving life of infants and the aged and are faced with unmanageable population increase and its attendant dangers, such as famine. We develop methods for the efficient production of commodities, but the repetitive and mechanical character of these production methods are such as to alienate and anger the working population and to render them ripe for rebellion. We devise cybernetic controls for operating productive machinery, then the working man is deprived of work. We provide conveniences and labor saving devices in abundance but in a bureaucratic organization and within a world view which deprives men of their freedom and dignity. We produce bread efficiently and thus remove the occasion for charity. In general, we behave like facile intelligences in infantile personalities, as our solutions to problems regularly generate worse problems. When the problems become unmanageable, we go to war or turn our major energies to devising mechanical means for escaping from the planet. And, indeed, nowhere has the extraordinary inventiveness and skill of technological man been so fruitfully manifested as in recognizing and resolving the difficulties encountered in whipping up whole nations to the fever pitch of hatred required to engage in modern war and then in prosecuting the war through to a conclusion against a powerful opponent.

The alternatives to these self-defeating schemes are unhappily limited; at the present impasse they seem to reduce to mass starvation, to some form of slavery, or to the further development of the cult of technology, which saves man's life while threatening his humanity. In a rather obvious sense the effects of our technology have done us violence. It will be well to consider, more closely, the precise sense in which damage has been done to the self, to others, and to nature.

We begin by considering the general character of violence. What is violence? In its general sense, I define it as treating a whole as if this whole were identical with one or some of its parts. In particular, violence offered to a person consists in behaving toward the person or self as if he were identical with some role or some special aspect of the self which is found to be interesting or which can be used. Thus, the criminal who mugs a passerby is acting out a partial view of the passerby, treating him as nothing more than an object which prevents access to the desired wallet. Pornography is a form of violence in that it ignores or finds valueless all aspects of a person except his sexual attributes. Similarly the investigator who persists in maintaining an objective attitude toward persons in order to play a fate-like role in studying or manipulating them in the interests of his curiosity or the unlimited Progress of science and technology is treating them as if they were identical with one of their attributes. He is, therefore, doing them violence.

Of course, we are again skirting an extremely difficult ontological problem: how can we behave toward or become aware of the whole being of anything? If we cannot, how can we avoid violence under any circumstance? Perhaps we cannot. Perhaps we cannot become fully aware of the self as a whole or as a fully concrete unity. Moreover, the conclusions we drew concerning the self were mainly negative, *viz.,* that it is dependent, finite, and not identifiable with any role. We are, though, constantly forced by circumstances into treating persons as functionaries or role-players of some sort. So soon as we assign a definite meaning or function to a self, we seem to be forced into treating him as just this single aspect or role. Thus, we tend inevitably to describe a person or to behave toward him as a physician, as a patient, as a citizen, as a student, or in some other standard and specialized capacity. The same observation holds true with respect to oneself, and also with respect to objects. Are we, then, condemned to do violence to anyone or to anything which requires our understanding or invites our concern? I believe this possibility should be taken seriously. Inevitably we react to a person or to an object as if each were identical with some trait or function. We treat the rabbit as a laboratory specimen, the janitor as the cleaning functionary. Theoretical or practical life could not be carried on otherwise.

Here I suggest we find ourselves again confronting what was previously termed the ontological fault. It will be recalled that the ontology sketched in section 4 (cf. also §7) specified that being becomes manifest to us only in a world, and a world opens up for us only a finite perspective. Some possibilities must always be hidden if others are to be exhibited. Some tasks must be left undone if others are to be completed. The consequence of these intrinsic limitations is that man is beset by the ontological defect (§15 and p. 108 f). That is to say, the worldly conditions under which a man is obliged to act are conditions which allow him only partial knowledge and deny him full achievement. Hence, self-misidentification and tragic error are inevitable. We cannot decide about, speak literally of, conceive precisely of, or behave toward the whole being of anything. We are limited in our dealings with anything to one aspect or finite group of aspects at a time. The consequence of this ontological defect is that the very conditions for forming a culture entail violence. Violence, however, takes different forms in different cultures. Here we distinguish three forms of violence: the first two are associated with humanistic and with technological cultures respectively. The third form arises from the relation — or lack of relation — between these two.

Dialectic of whatever sort aspires to be a self-corrective procedure. It is exemplified in work defined as activity which is controlled by its result. But this control develops by degrees. Any present human operation normally accepts as one of its factors a measure — a rough measure — of the degree to which the immediately past operation deviated from a pre-set goal. Then the present operation seeks to compensate for this deviation, but it proceeds inevitably in partial manner, a manner which requires further correction. We observed this kind of self-adjusting procedure when discussing the objective dialectic. We observed that an hypothesis concerning some natural event is tested and reformulated, then re-tested to accord ever more precisely with the data (cf. §37). Thus, scientific theories, when successful, become increasingly exact and extend to ever wider areas of experience. The technological use of these theories develops analogously. Here violence done to nature by imperfect and limited theories or unbalanced (e.g., un-ecological) uses may be supposed to be gradually corrected (or correctable).

In a not dissimilar way, we also noticed that the non-objective dialectic, following the cyclical form of the dramatic pattern, continually tends to bring the protagonist's insight back upon his own self-understanding, broadening and deepening it. Thus when successful, his interpretation of his own possibilities should unfold continually. We may, therefore, suppose that the violence which the protagonist does himself by reason of a merely partial grasp of his nature and his possibilities can, as this dialectic advances, be gradually corrected. The protagonist compensates for violence done himself by recapturing through insight previously ignored possibilities and values.

In addition, a third kind of violence results when either of

these two forms of dialectic is practiced to the (relative) exclusion of the other. When the objective dialectic is practiced
exclusively, then the violence of technism results. Similarly the
violence of humanism is the issue from a life devoted primarily
to practice of the non-objective dialectic. The violence of
technism arises from ignoring the self and avoiding its
characteristic pattern of development; the violence of
humanism follows from ignorance of nature. Both kinds of
dialectic and both kinds of violence are possible within the
structure of the Western world. Probably in any actual culture
belonging to this world, neither is absent. But there are differences in emphasis. A primarily humanistic culture is
deficient in knowledge of nature. Technological cultures tend
to minimize self-knowledge and the non-objective dialectic;
technism ignores them in principle. Violence arising from this
self-ignorance constitutes the threat of our time.

Technism by definition considers man to be in every respect a
proper part of nature and a use value. Such a man examines
himself only so far as he is a natural object like any other and
allows self change only within the cultural type or role which he
has adopted. According to the morality of technism, the non-
objective dialectic is a mere subjective diversion, the supposed
intrinsic value of all men is a piece of sentimentality, and the
attitude of loving care of nature is a bit of impractical
romanticism. The world-symbols of technism direct men,
instead, to regard themselves as workers for technology, at best
as brain-workers or problem-solvers. A popular expression of
the man of technism is the Cyborg, the man-machine; the man
who looks wholly to the machine for his salvation will see his
final form as the immediate union of man and machine in his
one person, as the Mediaeval man saw his ideal and complete
form in the unity of the divine and the human (cf. §27). And he
envisages others as co-servants of technism and nature as fuel;
all equally place their future in the hands of this prosthetic
savior.

Now this understanding of man, while efficient within its
limits, is clearly a partial understanding; accepting it as the

whole and sufficient view does man violence. And this violence is a peculiar one, for it is a self-destructive one. Since man under technism refuses to practice the non-objective dialectic, and since this dialectic is the means by which he might discover the partiality of his self-understanding, he will remain unable to see his way to compensating for this partiality. He is guilty of systematically ignoring the primary human obligation. And the inevitable consequence of this self-ignorance is self-destruction. The peculiarity of violence of technism, then, is its self-destructive character. We are reminded of Macbeth, who in the interest of acquiring power, murdered the King, source of light and structure in his world, only to find that by his deed he had annihilated himself as well.

In order to neutralize this inclination to violence, modern Western cultures face a two-fold problem: first to discover or to rediscover some effective form of the non-objective dialectic, second, to harmonize the exercise of this dialectic with the scientific and technological practices already so efficiently pursued. Our present technological culture is a compromise, for it has not yet lost all appreciation and grasp of the techniques of self-inquiry. It has not yet fully entered the world of technism. But it retains both the kinds of dialectic in an indefinite relation. I want to close with the reminder that this relation can be conceived more explicitly and in a form which conserves the basic instrumental function of technology and something at least of humanistic value. Ideally modern man could operate both dialectics. Then the fate-like role of the scientist-technologist, disposing of nature according to his own will, would be complemented by the interior dialectic of self-inquiry which determines a man's will in accord with his best insight into human possibilities and limits. Perhaps by way of such an exchange between the two dialectics, some compensation for the violence inherent in each may be achieved. As Plato might have put it: the human race will not see the end of its troubles until those who pursue the truth about themselves are also those who hold authority in disposing of technological power, and until those who design our technological society become by some divine commission seekers after the truth about themselves.

It may have appeared that this essay is technophobic. Not so. Not in that spirit did I remark in the Preface that the humanistic philosopher participates only ironically in his culture. These concluding reflections are intended to point to some possibility of compromise between the demands of technology and the needs of human life. Practice of the non-objective dialectic is not necessarily inconsistent with practice of the objective dialectic. But we have developed no world-symbols which point toward the effective integration of the non-objective dialectic into technological culture. We live, a philosopher has remarked, in a time of twilight, when the old gods have departed and the new are not yet come.

It is at least clear that a desirable harmony between the pursuit of self-knowledge and the tendencies of technology cannot be maintained so long as the two are conceived to be coordinate and competing specialties. Rather, they must be conceived to be Hierarchically related, the pursuit of self-knowledge being accepted as the indispensable prerequisite to technological specialization. In principle, the distinction can be maintained between technology as means and as end.[20] This distinction, however, can be retained and a utilization of technology as means can exist in a universe which preserves a recollection of the humanity of man only if we were to accept a doctrine of man not totally unlike that outlined in Chapters II through V together with the world-symbols which would insure its communicability and continued effectiveness. Were this common agreement to occur, then it might be possible to avoid slavery or famine without at the same time succumbing to a bureaucratic and technological organization which preserves society but only at the sacrifice of the individual's possibility of attaining courage and pursuing self-understanding within appropriate limits. The alternative to some such humanistic solution to the technological problem is surely the eventual violent destruction of man by man himself, the instrument of his suicide being the abstract space, time, nature, and man, so laboriously constructed.

# NOTES

## NOTES TO CHAPTER I

[1] In an earlier writing (*Philosophy at the Crossroads* (Baton Rouge, La.: University of Louisiana Press, 1972), p. 297f.) ) I suggested making a comparison between a critically purified experience and our scientifically disposed culture and by this means reaching a more secure understanding and evaluation of the scientific theories designed to explain experience and that experience itself. This essay does not move directly toward this end but rather approaches it obliquely. It will use the device of taking the measure of what I shall call scientific and technological culture.

[2] A philosophic exposition and defense of this view in several areas of thought is to be found in *Naturalism and the Human Spirit,* ed. Y. H. Krikorian (New York: Columbia University Press, 1944). The unity among the variety of naturalistic views is succinctly expressed by Professor Sidney Hook: it lies in "the wholehearted acceptance of scientific method as the only reliable way of reaching truths about the world of nature, society, and man," *op cit.,* p. 45.

[3] Some of the writers interested in relating the humanities to the cult of the sciences appear to me to exhibit an imperfect grasp of the humanities. Cf. C. P. Snow, *The Two Cultures: And A Second Look* (New York: Mentor Books, 1964). On the contrary, however, writings of the biologist, Jacob Bronowski, manifest a clear and sympathetic grasp of the nature and spirit of the humanistic tradition; cf. *The Identity of Man* (New York: Messner, 1956); *Science and Human Values* (New York: Messner, 1956); but his outlook is exceptional. At the opposite end of this spectrum is Castle, the representative of the humanities, in B. F. Skinner's best seller, *Walden II* (New York: McMillan, 1962); Castle is curiously without grasp of the tradition which his author wishes him to represent.

[4] "Man is the being whose relation to being is mediated by symbols," from "Homo Symbolicus" by Edward Henderson (*Man and World,* II, 1971), pp. 131-150.

[5] Cf. Chapter VI, note no. 9, p. 245.

[6] Victor C. Ferkiss, *Technological Man, the Myth and the Reality* (New York, N.Y.: Mentor, New American Library, Inc., 1970). He writes "Technological man will be in control of his own development within a context of a meaningful philosophy of the role of technology in human evolution" (p. 202, cf. also p. 68). Unfortunately, Ferkiss does not tell us either what this meaningful philosophy is or how nations will be induced to accept it.

Other books which place technological man and his culture in the role of Savior are these: Leon Trotsky, *Literature and Revolution* (New York: International Publishers, 1928); R.R. Landers, *Man's Place in the Dybosphere* (Englewood Cliffs, New York: Prentice-Hall, 1966); B.F. Skinner, *Beyond Freedom and Dignity* (New York: Knopf,

1971); Teilhard de Chardin, *The Phenomenon of Man,* trans. Bernard Wall (New York: Harper and Row, 1959); *The Future of Man,* trans. Norman Denny (New York: Harper and Row, 1965).

Some writings which reject the claims of technological messianism are the following: T. S. Eliot, *The Idea of a Christian Society* (New York: Harcourt Brace, 1949); *Notes Toward a Definition of Culture* (New York: Harcourt Brace, 1940); R. Hillegas, *The Future as Nightmare* (New York: Oxford University Press, 1967); Jacques Ellul, *The Technological Society,* trans. John Wilkinson (New York: Alfred A. Knopf, 1965); Martin Heidegger, "Die Frage nach der Technik" in *Vorräge und Aufsätze* (Pfullingen, Neske, 1954), pp. 13-44. Lewis Mumford, *The Transformations of Man* (New York: Harper and Row, 1956), Chapters 6 and 7; Jurgen Habermas, *Toward a Rational Society* trans. Jeremy J. Shapiro (Boston, Mass.: Beacon Press, 1970), see especially chapter 6; Herbert Marcuse, *One Dimensional Man* (Boston, Mass.: Beacon Press, 1964).

[7]Ferkiss, *op. cit.,* pp. 82-91; 202-223; and cf. below pp. 223f.

[8]Cf. *The Atlanta Constitution* September 14, 1974, p. 8A. For other instances of highly questionable experimentation upon human beings, cf. Daniel Labby, ed., *Life or Death: Ethics and Options* (Seattle: Univ. of Washington Press, 1968); also M.H. Pappworth, *Human Guinea Pigs: Experimentation on Man* (Boston; Beacon Press, 1968). Several aspects of this topic are discussed in an issue of *Daedalus* on "Ethical Aspects of Experimentation with Human Subjects," vol. 98, no. 2 (1969); a balanced and responsible view is expressed by William T. Blackstone in "The American Psychological Association Code of Ethics for Research Involving Human Participants: An Appraisal," *Southern Journal of Philosophy,* vol. XIII, no. 4 (1975), pp. 407-418.

[9]Hesiod tells one version of it in the *Theogony,* the Maoris of Polynesia tell another, (cf. J.C. Andersvn, *Myths and Legends of the Polynesians* (New York: Farrar and Reinhart, 1928), p. 367; the Zunis of North America have still another version, cf. F.H. Cushing, "Outlines of Zuni Creation Myths" in *Thirteenth Annual Report,* Bureau of Ethnology (Washington, D.C., 1896), pp. 379-82; and it occurs in *Genesis* (i, ii). F.M. Cornford comments on several of these and other versions and their relation to Milesian cosmology in his *Principium Sapentiae* (Cambridge University Press, 1952), chapters 11 and 12.

[10]Cornford points out (*Op. Cit.* chapter 11, p. 194) that Greek '*chaos*' — related to *chora* — meant empty place or room, e.g., the living space between heaven and earth.

[11]Compare the art described in Plato's *Politicus*; its aim is to weave into a harmony the aggressive warrior strain and the peaceable peasant strain, both in the state and in the individual soul (309). Of especial interest and importance in relating Greek history, religion, and philosophy is the article by Charles Bigger, "The Lesson of Apollo," *The Southern Review,* IX N.S., Spring 1973, pp. 334-358 to which I am much indebted.

[12]This view is developed in A.N. Whitehead's philosophy of organism. More recently, Hans Jonas has elaborated it; cf. his *The Phenomenon of Life* (New York: Harper and Row, Publishers, 1966), cf. pp. 38-62 and *passim.*

Perhaps I should add that a number of ontological problems cannot be considered in this essay, problems such as these: whether and in what sense the primary reality is monistic, dualistic, etc.; what the ontological status of possibility may be; what status is to be assigned to interpretation and interpretative principles; producing solutions to problems such as these, while interesting and important, will not materially affect the stand to be developed in this essay nor the conclusions to be drawn.

[13]I do not offer the preceding paragraphs as a sufficient theory of perception but as an indication of my use of terms. A sufficient theory would have, for example, to make distinctions among self (perceiver), presentation, and interpretation in a manner to

avoid the extra spectator fallacy which my grammar may seem to suggest (cf. note No. 4, p. 42). Were it developed less briefly, this view of perception would resemble Husserl's theory of active and passive genesis of meanings in the temporal flow of hyletic data (cf. his *Formal and Transcendental Logic,* tr. Dorian Cairns (The Hague: Nijhoff, 1969), pp. 314-21, also his *Cartesian Meditations,* tr. Dorion Cairns (The Hague: Nijhoff, 1960), ff37-39.

[14]The general pattern of these virtues is pointed out and illustrated in my *Socratic Ignorance* (The Hague: Nijhoff, 1965), pp. 57-66.

[15]Cited by Hans-Georg Gadamer, *Truth and Method,* tr. Garrett Barden and John Cumming (New York: The Seabury Press, 1975), p. 28; Gadamer offers considerable historical discussion, *ibid.*

[16]My expression suggests that there may be more than one world, but this supposition has often been denied. Both Aristotle and Kant deduced the singleness of the world, the all inclusive whole, from their basic principles. Husserl, on the other hand, did not deduce the world from prior principles. Rather, he discovered that he always already had a pervasive belief (a *Glaubensboden*) in the all embracing world, a whole always intuitively given as there (*Ideas,* §30; for bibliographical data see note 18 below). He also held that the notion of a second world, though not formally self-contradictory, is existentially incoherent. Like Husserl, I conceive of the world not as a thing but as the largest context of things. I discover it as the conviction which is related to all my other convictions; it is the encompassing context of my belief. Also it is a structured context, the horizon of horizons. Having a definite structure, the world is at any time a limited context; other such world-contexts seem to me — in contradistinction from Husserl — always to be possible, cf. §§7, 35 below. Heidegger also found the world to exhibit a structure. But the structure which he elaborates in *Being and Time* (§§15-18) is limited to work relations, to the producer's dealings with things of nature. The structure of a world, however, is both broader and more general than this set of relations, important though they are. For the world contains all beings and the relations among them. We may conceive it as that which contains and relates the three primary ways in which one experiences beings; thus, world is a form which relates my self, other persons, and the environing nature. The world thus conceived, however, is not unique. The way in which I structure self, other, and nature is not the only possible structure. Indeed it is empirically evident that there are many kinds of natural-world beliefs. For instance, there is every reason to suppose that the pre-literate Greeks and the Greeks of the Golden Age structured their worlds quite differently (cf. Eric Havelock, *Preface to Plato,* Cambridge, Mass.: Harvard University Press, 1963), and that the latter in turn did not live in the Christian Western world. Clearly there are both similarities and differences between these two kindred worlds. Only confusion, though, follows from failing to distinguish between the finite world of the polis in its cyclical time and the infinite world state in its linear time stretching toward some indistinct culmination. It will be useful in this essay to distinguish between these two worlds chiefly by reference to contrasting factors attributed to the self. Subjectivity, which is that through which one has a world and which is differently structured for different worlds, will be discussed in §§19 and 20 below. Also, it may be possible to place worlds in an order such that they point to world-in-itself, but such an order is not necessarily an historical order; cf. my "On Being and the Meaning of Being," *Southern Journal of Philosophy,* (IV. No. 4, 1966), pp. 248-65.

[17]The broadest definition of culture associates it with the totality of communicable items. But the world itself is a communicable item; hence, this way of defining culture confuses it with world. Another and more restricted way of defining it narrows the essential communicable items down to a definite number. One such definition lists something less than a hundred characteristics; it includes such matters as cosmogonic myths, religious rituals, marriage customs, and the like. More manageable still is the

following: culture "involves the proliferated use and manufacture of tools, the utilization of sources of energy outside the body (fire, for example), and a language which is independent of emotional reflexes and infinitely expressible from *Prisoners of Culture* by George A. Pettitt (New York: Chas. Scribner's Sons, 1970) p. 33, cf. p. 48. Undoubtedly language is the most important one among these three factors. Culture, then, might be defined as the form of life developed by the language using being. However, while these definitions are useful to remember, it will be more germane to our purposes to think of culture as the modes in which a world — a selection from the possible relations among self, others, and nature — are concretely and widely expressed. Culture is world as actually lived. It is a spiritual tool with which one fashions his being into the image of humanity which his world offers him. Western world will be described in §§17, 27, 38, and the family of cultures associated with it will be indicated on pp. 61ff, 235f, and 283-5; cf. also pp. 161-66.

[18]Op. cit., p. 206.

[19]"Our discussion will be adequate if it has as much clearness as the subject matter admits of, for precision is not to be sought for alike in all discussions," (*Ethics,* I, 3). I accept the injunction, taking it to recommend a methodological pluralism, and oppose the methodological monism of positivism which holds the scientific method is the only appropriate one for investigating any subject matter (Cf. note 1a, p. 41).

[20]Cf. his *Ideas,* trans. W. R. Boyce-Gibson (London: George Allen and Unwin, Ltd.; New York: The MacMillan Co., 1933), pp. 101-186; also *Cartesian Meditations,* Meditation I. Cf. also Chapter IV, note number 6, p. 1 below.

### NOTES TO CHAPTER II

[1]Erich Fromm has observed that in the contemporary Western world the competitive life of the market place so absorbs and dominates its participants that they tend to experience each other and even themselves only or primarily as saleable commodities; cf. his *Man for Himself* (New York and Toronto: Rinehart and Co., Inc., 1947), Chap. 3.

[2]According to Marxist doctrine, man evolved from simpler forms of organic life, the agent of this development being labor; there is no common human essence and hence no measure of man. Man, laboring to change nature, changes also himself. Thus, he is, at any historical epoch, the outcome of the dialectical interchange between himself and his environment. Friedrich Engels, in an essay, "The Part Played by Labor in the Transition from Ape to Man," (in *Dialectics of Nature,* trans. Clemens Dutt, New York: International Publishers, 1940), pp. 279-296, writes that the crucial event was the combination of circumstances in consequence of which man acquired erect posture, for then his hand was freed for the labor which both developed the hand and changed the world. Language was a later acquisition, cultivated to meet the practical need for cooperation necessary to carry on the more complex forms of labor which the developed human hand rendered possible. It is difficult to find in this doctrine any basis for the unfavorable measure and judgment of non-communist man which is associated with Marxists. For whether alienated or not, feudal man or capitalist man are, according to determinist (Marxian) doctrine, merely the products of their histories, no less and no more so than communist man. Of course, Marxian doctrine fails in principle to give due weight to the effect of self-awareness and of critical self-interpretation in man's development and to the dependence of these powers upon language.

[3]Several views of the self which differ from the present one will receive some passing notice or criticism in this essay. For convenience, I list their locations here. For remarks

on Aristotle's view of man, cf. this chapter, note 10, below, and chapter III, note 5 on page 241 on a determinist and naturalistic view, cf. §36, also this chapter, note 1, on p. 90; on a point made by some language analysts, cf. §24, on the Marxist view, cf. the note 2 immediately above.

[4]Reflective self-interpretation is not introspection; it is no less an activity of interpretation than is any other kind of perception. The distinction is important. The introspectionist view is an empiricist doctrine held by those philosophers who believe that all awareness is passive and that the mind can be engaged only in passively registering or in combining impressions of the senses. Those who picture mind in this fashion now argue that the mind does not peer at objects "inside" it as it peers at objects "outside" in the world. For mind has no "inside." Such spatial categories do not apply to it. Hence, there is no self inside the mind which one can observe and report on. Self-awareness, therefore, is an illusion. But the theory of mind as passive and as occupied in passively watching objects, whether outside or inside, seems to me to be completely at odds with experience and is explicitly rejected in this essay. The contrast between the two theories is drawn by Frederick J. Crosson in "The Concept of Mind and the Concept of Consciousness," in *Phenomenology in America,* ed. James M. Edie (Chicago, Ill.: Quadrangle Books, Inc., 1967), pp. 186-96.

[5]*Confessions of Saint Augustine,* tr. E.B. Pusey (New York: E.P. Dutton and Co., Inc., 1945), bk. II, chap. 3.

[6]St. Augustine writes: "And we indeed recognize in ourselves the image of God, that is of the supreme Trinity . . . for we both are and know that we are and take delight in our being and our knowledge of it." *The City of God,* tr. Marcus Dods (New York: Hafner Publ. Co., 1948), vol. I, p. 468. Thus, one's being, knowledge, and love constitute the image of the Trinity. Cf. p. 149 below.

[7]Cf. St. Bonaventura, *The Mind's Road to God,* tr. George Boas (The Liberal Arts Press, 1953).

[8]In order to evaluate the influences which have formed our present world, no doubt the distinction between Mediaeval Christianity, with its doctrine of the fall of man, and Renaissance Protestant Christianity, with its (Calvinist) doctrine of the total degradation of man, should be borne in mind. According to the first view, man is indefinitely removed from his completeness, yet he is redeemable with divine assistance. According to the second, he is infinitely remote from his completeness and requires nothing less than recreation—or abandonment of the whole religious conception of his being (cf. pp.113f, 224). The Mediaeval doctrine should also be contrasted with another more recent competitor: the Rousseauistic teaching which denies the fall and affirms the natural goodness of man. The origin of human evil is there discovered in defective institutional arrangements. Comteanism, Marxism, and various forms of scientific humanism adopt this view and reasonably conclude that man's salvation is eventually to be accomplished by secular and institutional changes.

[9]Here I do not use *logos,* language, in the narrow sense in which it is often understood in the definition: "Man is the language-using animal" but in the wider sense where it refers to any act of interpreting, any act of using one thing or part of experience to refer to another by way of some (subjective) intermediary. For illustration, cf. §20, also §19. Although in any world man is the self-interpreting being, still this power of *logos* manifests itself in characteristically different ways in different worlds.

[10]Aristotle held the human being to be a composite of matter and form, like other sub-celestial objects; only its form, rationality, was different in act from any other substance. A uniformity of view is thus obtained, but at the expense of justice to common experience. Cf. Chapter III, note number 5, on page 241f. On self as object in a modern sense of object, cf. §36.

[11]Stuart Hampshire in *Freedom of the Individual* (New York: Harper and Row, 1965), p. 118, draws a distinction between empirical and human possibility which parallels my distinction between objective possibility, with which I am only incidentally concerned in this chapter, and non-objective possibility.

[12]The "magic circle" is such a symbol. See M.L. von Franz, "The Process of Individuation," in *Man and His Symbols,* ed. C. G. Jung (New York, Garden City: Doubleday and Co., Inc., 1964), pp. 213-218.

[13]This last consideration points to the kind of personality problem which concerns the psychotherapist and which does not fall within the range of this essay. See Eugene T. Gendlin "A Theory of Personality Change" in *Personality Change,* eds. Philip Worchel and Donn Byrne (New York: John Wiley and Sons, 1964), pp. 100-148.

[14]I understand a dialectical relation to hold among a set of elements which are mutually causally related and goal-directed, so that the movement toward the goal is self-corrective. Examples: a mechanical system with feed-back; a Socratic conversation. This understanding of dialectic is elaborated further in my article, "On the Nature and Use of Dialectic," *Philos. of Science,* 22, no. 3 (July, 1955), 205-13. The particular kind of dialectic which I have in mind here is indicated by Martin Heidegger in *Being and Time,* Division II, chapter II; cf. also Stephan Strasser, *The Idea of Dialogal Phenomenology* (Pittsburgh, Pa.: Duquesne Univ. Press, 1969), Lectures 4 and 5; and cf. below, chapter III and §22f and 25, also p. 231.

[15]My purpose in this chapter will not be advanced by consideration of the familiar techniques of advertising nor of the more subtle forms of social pressure designed to force the development of character into channels favorable to the *status quo* or to some special interest.

[16]On the language utilized by these techniques and its contrast with a literal use of language, see pp. 183-87 and also, this Chapter, note 20 below.

[17]I use "person" as it seems often to be used, sometime to refer to the current self, sometimes to refer to the role-player. Etymologically the latter use seems clearly to be justified, for *per-sona (per-sonare)* refers to the mask through which the actor spoke or shouted his lines to the audience. But the other usage is common, too. My meaning will be evident from the context. I use "individual" when the distinction between self and role is not made.

[18]Cf. Erich Fromm, *The Sane Society* (New York: Holt, Rinehart and Winston, Inc., 1955), chaps. 2 and 3, and cf. William H. Whyte, Jr., *The Organization Man* (New York: Simon and Schuster, Inc., 1956), p. 6ff and *passim.* A usual trait of this social character is routine acquiescence in the local culture and myths.

[19]I elaborate this point in an article, "Individual and Person," *Philos. and Phen. Research,* XVIII (1957), pp. 59-61.

[20]The power, subtlety, and variety of metaphor and analogy are indicated by Philip Wheelwright in *The Burning Fountain, A Study in the Language of Symbolism* (Bloomington, Indiana: Indiana University Press, 1968), especially chapters 5 and 6; also by Paul Ricoeur, *La Metaphore Vive* (Paris: Editions du Seuil, 1975). See also the wise and witty essay by Walker Percy, "Metaphor as Mistake," in *The Message in the Bottle* (New York: Farrar, Straus, and Giroux, 1975), pp. 64-82. This topic is broached below in its relation to the languages of measurement, §3.

[21]I do not mean to suggest that the poet is somehow exempt from playing a role in society. If though, he does play a role, his role is quite exceptional, rather like that of the political revolutionary. Both are roles which are destructive of ordinary ones, of routines, of the entrenched order, and upon occasion may bring a culture or even world into question.

NOTES TO CHAPTER III

[1]Obligatory acts or valuations are often termed simply obligations. "Obligation" also names the relation among A, B, and C, once A, B, and C have been identified. This ambiguity need cause no difficulty once it is recognized.

[2]Value will be considered to be the satisfaction of felt need. Valuation is the interpretative act of assigning value or probable value to an object or event; the assignment may occur at both spontaneous and reflective levels of experience.

[3]I assume that anyone who is curious about obligation will be able to provide an ostensive definition of guilt, that sense of blameable deficiency in which the person himself is peculiarly involved. In the present writing I do not consider the positive correlative to guilt, the feeling of moral self-approval.

[4]Freud has speculated in an interesting psychological myth upon the initial act in the formation of conscience. He thinks of it as an actual act of (often repeated) cannibalism in which a group of the more ambitious brothers in the primitive horde banded together, driven to extremes by continual frustration of their sexual and other interests, by the powerful and godlike father (cf. his *Totem and Taboo*). They murdered and ate the father. This act was not the simple removal of a barrier to gratification. For the father was not only the powerful protector but, after a fashion, he was also loved and admired. Eating him was the way of acquiring his mana, of becoming the father, but only at the price of annihilating the loved figure and the tabooed embodiment of power and security. Thus, the sons incorporated the father and acquired power to satisfy appetites, but only at the expense of suffering guilt. This incorporation of the father and taking responsibility for themselves awakened ambivalent feelings. To speak of this attempt to become the father as the source of an original sin of mankind is to affirm an inevitable and universal development of a sense of guilt in the route to becoming a self which any man takes. There is good evidence that such a sense does tend to develop, whether Freud's myth points toward its inception or not. I allude to it later in terms of an ontological defect (§5).

[5]It should be observed that the present view differs from the classic view of man which holds that the good or mature self is reached by realizing those potentialities defined by his eternal and unchanging human essence. First note that this classic view denies the full impact of time and history. History really makes no difference to an unchanging essential humanity. The passage of time only provides the occasion for growing into just those powers compatible with or specified by the essence. Aristotle's view of this essence has been determining: "Reason is, in the highest sense, a man's self" (*Ethics,* X, 7). But reason has become more and more abstract and specialized. In modern cultures it has tended to be identified primarily with mathematical thinking. The notion of a rational essence, thus, became in fact a Procrustean measure. Another difficulty with this view of man's essence lies in the fact that a man's individuality, his personality, his particular virtues, are understood to be accidents of the essential self. They do not belong to what he primarily is. Hence, Socrates' wisdom and his courage, for example, for which he has been remembered for the last two and a half millenia, are not part of the essential man. The supposition, however, that history does not touch man's essence violates the sort of intuition of self embodied in the myths of man's fall and possibly exemplified upon other occasions of radical change. Also, to suppose man is in essence only his rationality or his rational animality fails to do justice to the manifold possibilities which men feel and experience in themselves or which are exemplified in other cultures. Finally, the conviction that a man's virtues or his concrete personality are not essentially his violates common experience of selfhood and convictions concerning the value of the individual. These reservations, however, do not extend to Aristotle's view that a man is the power to

choose to initiate action in accord with a principle of which he is aware (*Meta* IX, 4; *Ethics*, III, 2).

[6]Cf. my *Philosophy at the Crossroads,* §3, and §11, above.

[7]Note that the problem of reconciling humane studies and human values with scientific ideals and pursuits persists only in the context of the more recent modern culture. In the older Western cultures there are problems enough, but one matter it has settled: the relation of an impersonal knowledge of nature, the power, and the economic wealth which it produces, to specifically human values and obligations clearly ought to be the relation of means to end. Without this subordination, it conceived that no civilization, no self, is possible. Cf. Max Horkheimer, *Eclipse of Reason* (N.Y.: The Seabury Press, 1974), Chapter I.

## NOTES TO CHAPTER IV

[1]The structure of the term re-cognition is suggestive: "cognition" is just the seeing of a particular kind; the *recognition* is that seeing over again in other (sensed, felt, imagined, or conceived) elements. More generally, the possibility of a discovery anticipates its appearance. In the puzzle of experience and knowledge there was already a place for the new item and thus a clue to it.

[2]I do not mean to appear to accept Kant's theory of the cognitive self. The transcendental self, here alluded to, is a formal unity, a necessary presupposition of scientific knowledge. My discussion of the self (Chapter II and *passim*) is intended to point to the concrete self. Cf. my *Philos. at the Crossroads,* §28.

[3]This point is developed in my " 'Alms for Oblivion': An Essay on Objective Time and Experienced Time," in *Phenomenological Perspectives,* ed. Philip J. Bossert (The Hague: Nijhoff, 1975), pp. 168-87, and cf. pp. 180, 322f. below.

[4]For more recent discussions of this view, see Charles P. Bigger, *Participation, A Platonic Inquiry* (Baton Rouge, La.: Louisiana State University Press, 1968), Appendix A, p. 197 *et seq.*; William S. Weedon, "A Theory of Pointing," *Southern Journal of Philosophy,* I (1963), pp. 20-35.

[5]I note, in passing, a few other possibilities of man which are to be included in the notion of subjectivity. Important among man's possibilities is the power of being alone or of being with or empathizing with another, of playing at being another person, of being at one with oneself or being alienated, of loving or hating, of laughing or weeping, of being certain or uncertain of one's place in the world, of holding or rejecting certain myths, of doing mathematics, physics, or the like; "seeing" or "illumination" is essentially involved in all these possibilities.

[6]We approach, thus, through the analysis of a metaphor, the transcendental or originative subjectivity which Husserl approached through his epoche (cf. *Ideas,* part II; *Cartesian Meditations,* Meditation I). We shall not, however, be interested in a Husserlian analysis of this subjectivity but rather in the growth of a subjectivity to maturity and in the condition, direction, and boundaries of this growth.

[7]Cf. *Tim.* 37A-B and F. M. Cornford, *Plato's Cosmology,* (New York: The Humanities Press, Inc.; London: Routledge and Kegan Paul Ltd., 1956), p. 94; cf. also *Meno* 81C-D; *Phil.* 30A.

[8]Cf. Plato, *Theaetetus,* 152D-160C, especially 156D-E; *Rep.* 507. Professor Charles Bigger elucidates these suggestions in his *Participation, A Platonic Inquiry,* pp. 135-145.

[9]Erwin W. Straus, "The Forms of Spatiality," *Phenomenological Psychology*, trans. E. Eng (Basic Books, 1965); Herbert Spiegelberg, "On the Motility of the Ego," in *Conditio Humana*, to Erwin W. Straus on his 75th Birthday, ed. W. von Baeyer and R. M. Griffith (Berlin; New York; 1966), pp. 289-306.

[10]Perhaps someone may observe that in this section I seem rather to be describing the dialectic of the other (cf. p. 136). The self's relation to the other, however, is always a function of its relation to itself; thus, one must to some extent have become a self by way of the non-objective dialectic in order to be related as a self to another self. This section will emphasize the development of the self, although of course in its relation to the other.

[11]Cf. Frank Waters, *Masked Gods, Navaho and Pueblo Ceremonialism* (New York: Ballantine Books, Inc., 1970), part II, chap. 2.

[12]*Op. cit.*, p. xvii.

## NOTES TO CHAPTER V

[1]The view that the protagonist's decision initiates a series of connected events is not necessarily refuted by the determinist's objection which holds that all the movements a man makes are in point of scientific fact the (mechanical) product of inheritance, environment, and various physico-biological factors. The Aristotelian view, which I have just outlined, of the nature of the event initiating a drama is defined by its logical relation to succeeding events. Although the statement "I shall become a physicist," may be the conclusion of lengthy deliberation and various psychological and biological factors, still it may initiate a series of actions and events which are logically independent of this deliberation, of previous states of mind, and of other preceding events. Nevertheless, the determinist argument is sometimes thought to refute the common conviction that a man can make decisions. Hence, two remarks will be made in defense of the standpoing of this essay. (1) A decision is an event which initiates a group of events having a logical unity and a dramatic structure. From the coign of vantage of common experience, it is indisputable that men do make such decisions. The mechanist made one when he decided to be governed by the evidence as he saw it and to espouse mechanism. But of course, within other contexts, other analyses of the same experience may be equally valid. It is essential to keep the concrete context of common experience distinct from abstract contexts designed for special functions or to resolve special problems. The freedom exhibited in one context may without contradiction have been abstracted in an other. (2) Environment, culture, and to an increasing degree even inheritance, are affected by men. In our culture and time the physical environment is mostly man-made. To the extent that man is the creature of culture and environment, he is the product of that which he himself has made. At least in this secondary sense, then, man is self-determined. Thus, man can act to change himself, either directly through his own decisions or indirectly through his culture. For further considerations relative to mechanism, see §§30 and 36.

Finally, the well known paradoxical character of determinism must be recalled: if determinism is advanced as a philosophical theory, it is advanced as generally applicable; if it is generally applicable, it applies to the determinist's theory; if the determinist's theory is determined, the distinction between truth and falsity is not relevant to it; if this distinction is irrelevant, the theory cannot be decided to be either true or false.

[2]A kind of freedom, different from that mentioned in the preceding note, is required for achieving the end, the insight. This freedom is the freedom for seeing, for reaching a rational grasp of that which was projected in the decision and that which was achieved

in the action, and for comparing these with that vague vision of man as a whole and of his fate. Cf. §§ 8-20, and the end of the preceding note.

[3]I have sketched these and related matters in articles in the Dictionary of the *History of Ideas,* ed. Norbert Wiener, etc. (New York: Charles Scribner's Sons, 1973), see under Tragedy, vol. IV, pp. 411-17; and under Comedy, vol. I, pp. 467-470.

[4]"Inappropriate" might have two senses here: (1) it might refer to a skilled and ethical artist who, however, must practice his role in the context of a culture which systematically blocks at some point the non-objective dialectic in its movement toward the whole self; (2) or it might have reference to a person who is limited in his capacity to play roles, limited in ways of which he cannot but be initially ignorant; here the ontological defect is relevant; this defect becomes evident in the insight which terminates the dramatic movement. This second sense is the one to emphasize in the present context.

[5]*The Concept of Mind* (London: Hutchinson and Co., Ltd., 1949), Chap. I, §2.

[6]It will be observed that I am using the notion of unspecialized cells, those which contain the possibility of the whole organism, rather as I used the notion of the poetic word in §1. Both are metaphors for the whole self.

[7]*Memories, Dreams, Reflections,* recorded and edited by Aniela Jaffe, tr. Richard and Clara Winston (New York: Random House, 1963), pp. 32f. Jung dates the experience at about his twelfth year. A somewhat similar recognition is recorded by Dorothy Sayers; she writes: "I perfectly well recall the astounding moment when the realization broke in upon my infant mind that every other person in the world was an "I" to himself or herself, as I was to myself," *Begin Here* (New York: Harcourt, Brace and Co., 1922), p. 24. Also cf. Herbert Spiegelberg, "On the 'I-am-me' Experience in Childhood and Adolescence," *Rev. of Existential Psychology and Psychiatry,* IV (1964), 3-12.

[8]See the remarks on St. Augustine's interpretation of this doctrine, p. 58f and Chapter II, notes 6 and 8, p. 239. Cf. also §4D and 7.

[9]Plato provides another such world-symbol in the *Symposium* (190f. ). He has Aristophanes give a playful account of the aboriginal double human being, who in the pride of his power, challenged the Olympian gods. In punishment for this insolence, the original man was split down the middle into two incomplete beings, each doomed to seek after that which would complete him, his lost human half. Socrates then seems to open a vista toward that which would complete the soul's deficiencies by giving an account of the movement toward an intellectual grasp of the Idea of Beauty. But a drunken brawl then ensues which suggests, with a full measure of Platonic irony, that such an intellectual apprehension of the principles of order and beauty is a doubtfully effective means for approaching the perfection of being, for it does not make contact with an integrate the a-rational part of man's being.

[10]An important difference between Mediaeval and Renaissance Protestant interpretations of the Fall of Man is mentioned in Chapter II, note no. 8, p. 239.

[11]Cf. William Gilman, *Science: U.S.A.* (N.Y.: Viking Press, 1965), pp. 281ff.; and Daniel Bell, editor, *Toward the Year 2000: Work in Progress* (Boston, Mass., 1969), especially the essay by Herman Kahn and Anthony J. Wiener, "The Next Thirty Three Years: A Framework for Speculation"; Kahn and Wiener anticipate the "practical use of direct electronic communication with and stimulation of the brain," p. 80f; see also Alvin Toffler, *Future Shock* (New York: Random House, Inc., 1970), pp. 209-15.

[12]A philosophical consideration of this kind is developed by Edward H. Henderson in "The Christian Transformation of the Ritual Way," *Anglican Theological Review,* April, 1973, pp. 189-200.

NOTES TO CHAPTER VI

[1]The history of instruments for the measurement of weight, temperature, voltage, etc., illustrate this dialectical development.

[2]The contrast between the space and time of the Mediaeval age and the space and time of the post-Cartesian era is in this respect very instructive; I have developed some aspects of this contrast in my *Philosophy at the Crossroads,* cf. § 4 and chap. III.

[3]Of course later views infer from the Heisenberg indeterminacy principle that there is a theoretical limit to the possibilities of exactitude in measurement; cf. R. Furth, "The Limits of Measurement," *Scientific American,* 255, July, 1950, pp. 3-6.

[4]Nor does the physio-psychologist capture and communicate this anxiety and excitement by measurements of changes in skin temeprature, pulse rate, respiration, etc. He may assume that changes in these quantities are somehow correlated with the ways in which the situation was experienced, but the experience itself is non-quantitative and escapes his techniques. The fine artist, though, e.g., the dramatist, might communicate such an experience by mimetic methods.

[5]C. N. Campbell, *An Account of the Principles of Measurement and Calculation* (London: Longmans, Green and Co., 1928).

[6]This relationship has been described in another way: the variables occurring in the equations of physical theory refer not to numbers but to pairs; one member of such a pair is a number, the other is a physical entity or property. Measurement, then, is the technique of selecting the number which is paired off with a physical entity or property. Cf. Karl Menger, "On Variables in Mathematics and Natural Sciences," *Brit. J. for the Philos of Sc.,* no. 18, 1954, pp. 135; and Peter Caws, "Definition and Measurement in Physics" in *Measurement, Definition and Theories,* ed. C. West Churchman and Philburn Ratoosh (New York: John Wiley and Sons, 1962), p. 11.

[7]More specifically, the rules for operating with these numbers specify association, distribution, and the inverse and unity operations for additions and multiplication; sometimes commutation and monotony are also specified.
On affine numbers, used in the measurement of equal intervals cf. Lynn N. Loomis and Shlomo Sternberg, *Advanced Calculus* (Reading, Mass.: Addison-Wesley Publ. Co., 1968), pp. 40f, 52ff.

[8]"Operational definitions (e.g., of measurement), in spite of their precision, are in application without significance unless the situations to which they are applied are sufficiently developed so that at least two methods are known of getting to the terminus." "Symposium on Operationalism," *Psycho. Rev.,* V, 52, 1945.

[7] [9]Cf. H. M. Johnson, "Pseudo Mathematics in the Mental and Social Sciences," *Amer. J. of Psycho.,* April, 1936, pp. 343-51; J. Guild, "Are Sensation Intensities Measurable?" "Symposium on Measurement," in *Rept. British Assoc. for Adv. of Science,* V, 108, 1939, pp. 296ff.; C. Sparrow, "Measurement in Social Science," in *Voyages and Cargoes* (Richmond, Va., 1947), pp. 150-179; also my article, "Operational Definitions and Theory of Measurement, *Methodos,"* 1953, pp. 233-49. On the other hand, some writers have sought ways to legitimize these suspect measurements, or some of them, by enlarging the theory of measurement to include other procedures; cf. S.S. Stevens, "Mathematics, Measurement, and Psychophysics," in *Handbook of Experimental Psychology,* ed. S.S. Stevens (John Wiley and Sons, Inc., New York and London, 3rd ed., 1960), pp. 1-49.

[10]Cf. Iredel Jenkins, "Postulate of an Impoverished Reality," *J. of Philos.,* 39, 1942, pp. 533-47.

[11]There are, of course, many other kinds of numbers—irrationals, reals, complex numbers, transcendentals, quaternions, and others—but the four I have discussed are the ones used in the familiar kinds of measurement.

[12]And one not without its obscurities; see, for example, my "The Paradox of Measurement," *The Philosophy of Science,* 16, no. 2 (1949), pp. 134-6.

[13]To count a group of concrete and unique individuals, we substitute for each a standard individual and then count standard individuals. Now, we should have no difficulty in reversing this substitution process. The cardinal number assigned to the group may then be regarded as the number of unique individuals in the group.

[14]*"Das Ding"* in *Vorträge und Aufsätze* (Pfullnigen: Neske, 1954), pp. 163-181.


## NOTES TO CHAPTER VII

[1]It is to be noted, however, that Jacques Ellul offers reasons for the conviction that modern Western culture has become technological in its essence; *op. cit.,* 127 ff.

[2]Cf. Yale Brozon, *Automation: Impact of Technological Change* (Washington, D.C.: American Enterprise Institute for Public Policy Research, 1963); John Rose, *Automation: Its Uses and Consequences,* (Edinburgh and London: Oliver and Boyd, 1967), chapters 2 and 3.

[3]Ellul, *op. cit.,* Chapter II.

[4]I use "Progress" (with a capital) to refer to a derivative from the view of time and history which the Christians (Cf. St. Augustine, *City of God*) opposed to the Greek view of cyclical time and repetitive history. The Christians understood time to begin with the creation and to move in a retrograde course until its turning point when help to man came in the form of the Incarnation of the Second Person; thence, it was to move triumphantly to the salvation of man in the Second Coming. "Progress" refers to a secularized version of this history. It sees the evolution of animal life developing through man in haphazard and uncontrolled fashion until the beginning of the scientific age. This history reaches a turning point when, with the widespread application of science to technology during the Industrial Revolution, man acquired the power of controlling his environment and even his own evolution. Thence it is to move triumphantly to a golden age of peace, plenty, comfort, and leisure for all—an earthly Paradise. Concomitantly, naturalistic beliefs (cf. Chapter I, note 2, p. 235 and pp. 203-6 below) replace humanistic ones.

[5]In fact, the relation of a society to modern technology is already being treated as a technological matter subject to efficient manipulation. Cf. Jay W. Forrester, *World Dynamics* (Cambridge, Mass.: Alden Press, 1971). Of course, such efforts to comprehend technology technologically are valuable, yet only so long as they understand their own limits. Among the several variables that enter into Forrester's computer program—industrial production, etc.—one is entitled "quality of life." Forrester seems to believe it obvious that this variable is to be defined in hedonistic terms. With so simplified and inadequate a view of the quality of human life, it is doubtful that the limits upon the validity of his results can be made evident. On the other hand, a more sophisticated and adequate concept of the quality of life would probably not be quantifiable.

[6]Cf. Heidegger, "Die Frage nach der Tecknik," in *Vorträge und Aufsätze,* pp. 13-44. I offer an interpretation of Heidegger's view in Chapter IX of my *Philosophy at the Crossroads.*

[7]Happily many of the proposed techniques for modifying human behavior to suit the designs of an experimenter have so far achieved only modest success; cf. Gardner c. Quarton, "Deliberate Attempts to Control Human Behavior and Modify Personality," *Daedalus,* 96 (1967), 837-53. No less happily there is evidence that some scientists are coming to be aware of the need for the control of certain kinds of scientific activity. At a meeting in Davos, Switzerland in 1974 of The Conference on Applications and Limitations of Genetic Engineering: the Ethical Implications, Professor Paul Berg of Stanford University spoke of the dangers inherent in genetic engineering and called for a moratorium on experimentation of this kind until effective safeguards could be developed. However, he found little support for his views. Cf. Harvey Wheeler, "The Regulation of Scientists," *Center Magazine,* VIII, Nov. 1, 1974, pp. 73-77. More recent efforts, however, in this direction have been more successful.

A different kind of danger lurks in the genetic studies which are related to human life. Joseph Fletcher has described this kind of genetic engineering as aiming to "to control people's initial genetic design and constitution — their genotypes — by gene surgery (transduction) and by genetic design (insertion and deletion)," in *The Ethics of Genetic Control* (Garden City, New York: Doubleday, 1974), p. 56. One can scarcely quarrel with geneticists' interest in eliminating the birth defects and weaknesses which would hamper a person in all his activities. But some geneticists are also interested in improving the "quality" of infants. Here their efforts are directed by ideals which are moulded, perhaps naturally, upon themselves. As Fletcher remarks, "Quality control in birth technology will have to aim at selecting for intelligence and, where possible, lifting it," (*ibid.,* p. 75). Fletcher leaves no doubt that he has in mind the sort of Cartesian ratiocinative intelligence which makes a good professional biologist and a contributing member of technological cuulture. But it is well to remember that there are other forms of intelligence, well also to remember the fate of the dinosaur.

[8]The kind of corporate organization and regimentation to which I allude appeals strongly to men like B.F. Skinner; cf. his *Beyond Freedom and Dignity,* chapter 1 and passim. One would expect this regimentation to produce conformity and uniformity, an expectation borne out by Skinner's novel, *Walden II* (New York: Macmillan Co., 1948).

[9]On the comparability of human thought to computer results, see Donald G. Fink, *Computers and the Human Mind* (Garden City, N.Y.: Doubleday and Co., Inc., 1966), especially chapters 10-13; also cf. M.L. Minsky, "Artificial Intelligence," *Scientific American,* 215, Sept. 1966, pp. 246-260; H.L. Dreyfus, *What Computers Can't Do: A Critique of Artificial Reason* (New York: Harper and Row, 1973; reviewed by Bernard Williams, "How Smart are Computers," *New York Review of Books,* xx, no. 8, Nov. 15, 1973, pp. 36-40;" H.L. Dreyfus, "A Critique of Artificial Reason," *Thought,* vol. 43, no. 171 (1968), pp. 507-522."

[10]"Meet Shaky, the First Electronic Person," B. Darrach. *Scientific American,* 1973.

[11]"Computer simulated intelligence" is often distinguished from artificial intelligence in that the former is specifically intended to imitate human cognitive activity as closely as possible. Discussions of research on this topic can be found in H. Dreyfus, *Alchemy and Artificial Intelligence* (Santa Monica, Calif.: Rand Publ. P-3244, 1965); K. Sayre, *Recognition, A Study in the Philosophy of Artificial Intelligence* (Notre Dame, Inc.: The University of Notre Dame, 1965); Hao Wang, *From Mathematics to Philosophy* (New York: Humanities Press, 1974), pp. 283-87, 301-11. Also cf. note 12 below.

[12]Some of the contributors to this debate are the following: L.J. Cohen, "Can There be Artificial Minds?" *Analysis,* 16, 1955, pp. 36-41; Paul Ziff, "The Feelings of Robots," *Analysis,* 19, 1959, pp. 64-68; R. Puccetti, "On Thinking Machines and Feeling Machines," *Brit. Journal for the Philos. of Science,* 18, 1967, 39-51; K. Gunderson, *Mentality and Machines* (Garden City, N.Y.: Doubleday and Co., 1971) especially chap. 5; Bertram Raphae, *The Thinking Computer: Mind Inside Matter* (San Francisco: W.M. Freeman & Co., 1976).

[13]Gunderson, *op. cit.,* pp. 144-157, makes this point persuasively.

[14]Worth a comment, perhaps, is the science fiction movie "2001," which told the story of an exceedingly complex computer, "Hal." Hal was so complex and his program so clever that he conceived hostility for his operator, and thus he was motivated to eliminate his operator. This feeling and the decision which interpreted it seemed to be the consequence of the computer's great complexity and, of course, of its very clever programming. Notice that the computer's decision freed the computer of its original (mechanical) program. Notice also that to suppose mere increase in complexity and in programming frees a mechanism of mechanism is a confusion analogous to a once popular belief about infinite series in mathematics. This belief held that a converging infinite series would actually reach its limit, e.g., zero, because the last step became so very small that it ceased to exist (i.e., "became zero"). Sophisticated circuitry and astute programming may make computers converge toward intelligent results or human-like behavior (I refer to the kind of process or behavior which a machine can imitate). But to suppose the limit actually to be reached and the machine actually to become human would suggest making an irrational leap, a suggestion requiring a loss of sophistication comparable to the popular mathematical belief just mentioned.

The difficulty may also be judged to be an instance of the fallacy called misplaced concreteness. The fallacy arises from identifying a theory not so much with the data which it explains as with the concrete objects or experiences from which the data were derived.

[15]Cf. Jon von Neuman, *The Computer and the Brain* (New Haven: Yale University Press, 1958).

[16]Some of the plays and episodes which have occurred so far in this dramatic development are set forth by Thomas Kuhn in his excellent book, *The Structure of Scientific Revolutions* (Chicago: University of Chicago Press, 2nd ed. 1970).

[17]For a concise historical account of alienation as used in the present sense, see Albert W. Levi, "Existentialism and the Alienation of Man" in *Phenomenology and Existentialism,* ed. Edward N. Lee and Maurice Mandelbaum (Baltimore, Md.: Johns Hopkins Press, 1967), pp. 243-265.

[18]It may appear that I have tried to push the scientist into holding a contradiction, *viz.,* that he is, as a man, subject to fate or determination, which he, as embodiment of fate, arranges or disposes himself. It must be recalled, though, that we are dealing with the manner in which the scientist experiences his own activity in an ideally technological culture. Here there is no question but that science in the interest of Progress is the final interpreter of self-experience. In the actual world, however, the scientist reflects the actual world view in which he is bred; he reflects it in ways which do not often reach critical awareness in the minds of those who participate whole-heartedly in them. These ways may, therefore, be mutually opposed.

[19]By Eugen Herrigel, tr. by R. F. C. Hull (New York: Pantheon Books, 1953).

[20]Cf. Eric Weil, "Science in Modern Culture, Or the Meaning of Meaninglessness," *Daedalus,* 95 (1965), 171-89.

# INDEX